I0109871

HURRICANE KATRINA
IN TRANSATLANTIC PERSPECTIVE

EDITED BY ROMAIN HURET
AND RANDY J. SPARKS

HURRICANE KATRINA IN TRANSATLANTIC PERSPECTIVE

Louisiana State University Press)|(*Baton Rouge*

Published by Louisiana State University Press
Copyright © 2014 by Louisiana State University Press
All rights reserved
Manufactured in the United States of America
LSU Press Paperback Original
First printing

Designer: Michelle A. Neustrom
Typeface: Corda
Printer and binder: Maple Press

Library of Congress Cataloging-in-Publication Data are available
at the Library of Congress.

ISBN 978-0-8071-5843-2 (pbk.: alk. paper) — ISBN 978-0-8071-5844-9 (pdf) —
ISBN 978-0-8071-5845-6 (epub) — ISBN 978-0-8071-5846-3 (mobi)

The paper in this book meets the guidelines for permanence
and durability of the Committee on Production Guidelines for
Book Longevity of the Council on Library Resources. ∞

CONTENTS

vii   Acknowledgments

1   Introduction
Romain Huret

8   "Two Centuries of Paradox"
*The Geography of New Orleans's African American Population,
from Antebellum to Postdiluvian Times*
Richard Campanella

38   Explaining the Unexplainable
Hurricane Katrina, FEMA, and the Bush Administration
Romain Huret

50   Picturing the Catastrophe
News Photographs in the First Weeks after Katrina
Jean Kempf

70   "Wilt Thou Judge the Bloody City? Yea, Thou Shalt Show Her
All Her Abominations"
Hurricane Katrina as a Providential Catastrophe
James Boyden

81   Naturalizing Disaster
*Neoliberalism, Cultural Racism, and Depoliticization
in the Era of Katrina*
Andrew Diamond

100   Reformers, Preservationists, Patients, and Planners
       *Embodied Histories and Charitable Populism in the Post-Disaster
       Controversy over a Public Hospital*
       Anne M. Lovell

121   The Political Economy of Invisibility in
       Twenty-First-Century New Orleans
       Security, Hospitality, and the Post-Disaster City
       Thomas Jessen Adams

137   Faith, Hip-Hop, and Charity
       Brass-Band Morphology in Post-Katrina New Orleans
       Bruce Boyd Raeburn

153   Memory Lives in New Orleans
       The Process and Politics of Commemoration
       Sara Le Menestrel

178   Why Mardi Gras Matters
       Randy J. Sparks

199   Contributors

ACKNOWLEDGMENTS

This book began in December 2005 with a conference organized in Paris by Sara Le Menestrel. Only a few weeks after Hurricane Katrina hit New Orleans, French and American scholars tried to bring some rationality to such a moving and disturbing event. For launching the project, we are very grateful to Sara's initiative and the people at the École des Hautes Études en Sciences Sociales in Paris who helped organize the conference. In particular, we would like to thank Cécile Vidal and François Weil. Later, with Jim Boyden we presented some of our conclusions at conferences of the American Historical Association, the Organization of American Historians, and the Southern Historical Association. We are grateful to all participants in our sessions, including Adam Fairclough, Alecia Long, and Matthew Mulcahy.

To commemorate the fifth anniversary of Hurricane Katrina, a group of scholars from Tulane and France gathered for a conference in New Orleans from October 21 to 23, 2010. Six scholars from various departments at Tulane and six scholars from France presented papers, exploring how academics on both sides of the Atlantic viewed the hurricane and its aftermath. That conference was cosponsored by the Tulane Department of History and the Centre d'Études Nord-Américaines at the EHESS. Additional support was provided by the New Orleans Gulf South Center at Tulane, the Murphy Institute of Political Economy at Tulane, the Institut universitaire de France, and by the Florence Gould Foundation, an American foundation devoted to French-American exchange and amity.

In 2011, we organized a follow-up workshop in Lyon and Paris. Many people provided crucial support, especially Camille Amat, Brigitte Esnault, Nicolas Larchet, Vincent Michelot, Sandrine Revet, Alexandre Rios-Bordes, Evelyne Thévenard, and Jean-Claude Zancharini. We are grateful for the support of these individuals and organizations, without whom this book project would not have been possible.

HURRICANE KATRINA
IN TRANSATLANTIC PERSPECTIVE

Introduction

ROMAIN HURET

You can't afford to have a hurricane when you're earning seven or eight dollars a day.
   —Zora Neale Hurston, *Their Eyes Were Watching God* (1937)

"I understood that being poor meant not being a citizen," one New Orleans resident bitterly observed on the day after Hurricane Katrina devastated the city in late August 2005. Branded refugees by the American government and suspected of committing the darkest of crimes during the long week in which the city was cut off from the rest of the world, the poor seized the opportunity to tell their harrowing tales of days surrounded by water, weeks of wandering in search of shelter, and months of wondering whether they would ever be able to return home. For many Americans, Hurricane Katrina marked the moment in which they "rediscovered" poverty and social injustice in America, and eventually wondered about their social contract. Far from the nation's opulent suburbs, from New York's Fifth Avenue and California's Sunset Boulevard, the lines of people at the Superdome, the stadium where the poor took shelter during the storm, revealed the existence of Americans with little cash in their pockets, living paycheck to paycheck on the earnings from menial, insecure, poorly paid jobs. Journalists and politicians harped on the same question: how was such a disaster possible *in America,* the world's greatest superpower?[1]

   Although social scientists had written copiously about the deterioration of a range of social indices in the wake of welfare cuts and economic deregulation, the disaster ripped the veil off America's hidden poverty and displayed it to the world without the softening filter of cultural mediators. Toothless old women, homeless people, aimless youths, and addicts in search of a fix appeared on television in scenes resembling those described by William T. Vollmann, a writer with a talent for portraying the impoverished America of "poor white trash" and the mostly African American "underclass" of the ghettos.

1

In France, where links with Louisiana are old, many people expressed their concerns for the people of New Orleans and especially the musicians. In the left-wing newspaper *Libération,* the journalist Patrick Sabatier evoked a "city under water, which will remain a ghost town for weeks." Like all other commentators, he couldn't help interpreting the event in terms of the geopolitical context and highlighted "the sudden collapse of a wealthy and over developed society." Concerts to support artists in the city were spontaneously organized in Paris. People of New Orleans humorously thanked the French by wearing the sticker *Chirac rachète nous* ("Buy us back, Chirac") during Mardi Gras.[2]

The Crescent City as Symbol

Because of its social and ethnic makeup, and due to the media's construction of the event, New Orleans after Katrina offered an exceptionally concentrated image of discrimination in America, compounded by the city's bad reputation. Thanks to the media, the city's Lower Ninth Ward quickly came to symbolize exclusion at the heart of American prosperity. As in many other urban neighborhoods across the United States, the situation in the Lower Ninth has deteriorated considerably since the 1950s owing to a combination of unfavorable economic, migratory, and social factors at both the local and national levels. Residents of New Orleans with the highest incomes left to live in the suburbs, state aid dried up, and the recession of the 1970s left public finances in a shambles for decades. New Orleans is also exceptional because of its large African American population, which has been hit harder by poverty and exclusion than other ethnic groups. Thanks to tourism and the Mardi Gras industry, the city did grow, but the poverty rate remains quite high in this majority black city.[3]

Many commentators asked simple questions in the aftermath of Katrina: Why didn't people of New Orleans leave when they heard about the strength of the hurricane and the issuance of an evacuation order? Why didn't they heed the warnings broadcast by the media? Were they simply incredulous? A study conducted in a Texas shelter for late evacuees gives a partial answer to this question: Fewer than a third said that they had underestimated the strength of the storm. Many mentioned the difficulty of leaving, the habit of riding out hurricanes, the desire not to abandon homes and pets, and, last but not least, their confidence in the ability of the city's levees to protect them. But for the vast majority of people who sought shelter in the Superdome or stayed in their homes, their reasons for not leaving the city were related to the social environment. First of all, in order to leave, one had to own a vehicle, but tens of thousands of residents did not. Even if the public bus system had operated normally, it would have been able to evacuate only 10 percent of the population. This alone accounts for the decision of roughly a third of the city's people to remain in New Orleans. For others, the cost of gasoline and lodging in a motel were sufficient deterrents. In contrast to middle-class people, relatively few of the poor had family and friends outside the area, and this made evacuation more difficult. Without money or family to go to, the city's poor therefore chose to ride out the storm while protecting themselves against flooding with whatever means were at hand. In the flooded city, the poor became visible to the naked eye—for once, attracting the attention of the media and politicians, recognized by their fellow citizens, and granted a presence in the public arena.[4]

Katrina as a Singular Moment

In this respect, the documentary *Trouble the Water* (2008) is revealing. Kimberly Roberts, an African American from the Ninth Ward, filmed the disarray as people awaited the storm and built makeshift protections against flooding. Roberts is an authentic product of the neighborhood, whose population has attracted a good deal of interest among sociologists. Her mother died, and her father abandoned her, leaving her to be raised by her grandmother. She earns her living by doing small jobs and engaging in some shady trafficking with her husband and dreams of becoming a rap singer. When many of her neighbors went to the Superdome, she stayed behind, walking the familiar streets, bantering with her neighbors, and making fun of their solidarity in the face of despair—all with camera in hand. The city became a ghost town in which only the poorest and most marginal remained, along with the bodies of the dead. On the day after the storm, as the waters began to recede, these people received little help from the National Guard, which refused to allow them to enter a secure naval facility. Evacuation was now mandatory, but riding in the back of a truck proved to be difficult. Along with her companions in misfortune, Roberts made her way to northern Louisiana, where she continued to film, protecting her cassettes with garbage bags. There, the situation was no less dramatic or chaotic. The Federal Emergency Management Agency (FEMA) did little to help people who had lost everything in the flooding. Without any didactic or aesthetic pretensions, Roberts recorded the stories of people whom America had left behind. A mother rails against the U.S. Army for sending the poor to Iraq but not helping them at home. She refuses to allow her son to enlist. Later, a grandmother asks what has become of the United States. Roberts's film is exceptional because it shows what sociologists and historians often have a hard time describing. Poverty is more than just "doing without." It is also "attachment" to a place, no matter how dilapidated and uncomfortable, along with a social network that may be limited in scope but nevertheless remains invaluable. Thus we see that a large part of the explanation for people's refusal to leave has to do with their roots in a community with its own codes, language, and rules, which everyone accepts. In leaving the Lower Ninth Ward, Kimberly and her companions in misfortune lost what little they had managed to put together for themselves—in short, everything. Local social networks and the social authority that went with them fell apart when people evacuated: small-time drug dealers lost their clients, mothers lost their day-care providers, workers lost the jobs for which they were paid under the table. The energy of this small group of people is astonishing given the trials they had endured. But we can understand it, because we can see beyond the usual clichés about poverty and discrimination. The documentary managed to shed light on the intricacy of local tensions and national failures as well as to present an awkward feeling of loneliness and despair.

The Disaster as a Social and Political Construction

In December 2005, only a few months after the disaster, most of the authors of this book met in Paris to understand the event and go beyond oversimplified explanations. The conference, entitled "Louisiana Adrift" and organized by Sara Le Menestrel, was full of emotions and questions. Why did the Bush administration take so many days to intervene? What happened with the levees system? Where are the people of the Crescent City? Will they ever come back? Was Katrina really an extraordinary event? Doesn't it call to mind familiar representations of the poor, African Americans, or musicians? None of the participants were experts in the field of disasters; they brought their own fields of research in anthropology, geography, history, or political science to understand what had happened on the streets of New Orleans. The presentations, written during the emergency, were the first drafts of this book. As scholars were both American and French, views and interpretations often diverged about the role of religion, the place of the federal government, the process of memory, and the exceptionalism surrounding studies and analysis of New Orleans. These opposite views gave a rich fiber to the debates, and the following pages present the result of our discussions.

As many studies have shown, there is no such thing as a "natural" disaster. Far from being a benign phenomenon in the United States, disasters are part of the environment of millions of Americans since the time of the first colonies. Historian Kevin Rozario has perfectly described the strength of the culture of calamity in American culture. Moreover, disasters always reflect larger social, political, and racial trends. Acts of God are man-made and shouldn't be analyzed without such larger contexts. Tapping such a methodological framework, the book combines multidisciplinary approaches revolving around three main themes: the event itself, its racial and ideological dimension, and its legacy.[5]

First, the "long week of Katrina" raises questions about the political rationality of the Bush administration and the levee policy. Many studies have criticized the failure and the impotence of national and local decision-makers.[6] In return, politicians have tried to vindicate their actions. President George W. Bush in his book *Decision Points* (2010) contends that he "should have recognized the deficiencies sooner and intervened faster. The problem was not that I made the wrong decisions. It was that I took too long to decide." The essays in this book deemphasize the role of individual human beings and stress more structural explanations strongly rooted in the geographical, political, and social transformations of the city and the country.[7]

Second, as soon as the hurricane hit the Gulf Coast, racial debates ran high and peaked with Kanye West's motto "George Bush doesn't care about black people." On television, African Americans were branded as refugees and conservative pundits accused poor people of being responsible for their fate. Conservative intellectual Robert Tracinsky indicted the welfare state: "What Hurricane Katrina exposed was the psychological consequences of the welfare state. What we consider 'normal' behavior in an emergency is behavior that is normal for people who have values and take the responsibility to pursue and protect them. People with values respond to a disaster by fighting against it and doing whatever it takes to overcome the difficulties they face. They don't sit around and complain that the government hasn't taken care of them. And they don't use the chaos of a disaster as an opportunity to prey on their fellow men." Normative definitions of "survivors" and "victims" in the press revolved around racial stereotypes and underlined the civic virtue of Cajuns against the selfishness and lawlessness of African Americans. The intensity of such debates derived from the spreading of neoliberalism and cultural racism in twenty-first-century America.[8]

Third, every disaster gave birth to narratives which paved the way to individual resilience and collective reconstruction. The violence of the hurricane sparked many debates about the reconstruction of the city and the memory of the event. New Orleans became a series of contested landscapes by various social and professional organizations. Traditional civic ritual such as the second lines and Mardi Gras took on a new meaning as they deployed a new sense of identity and citizenship.[9]

Notes

1. Quotation in the documentary *Trouble the Water* (Elsewhere Films, 2008); Michael Eric Dyson, *Come Hell or High Water: Hurricane Katrina and the Color of Disaster* (New York: Basic Civitas, 2005).

2. Martin Gilens, *Why Americans Hate Welfare: Race, Media, and the Politics of Antipoverty Policy* (Chicago: University of Chicago Press, 1999); Michael B. Katz, *The Price of Citizenship: Redefining the American Welfare State* (New York: Metropolitan Books, 2001); Patrick Sabatier, "Editorial," *Libération,* Sept. 1, 2005, 6.

3. See Nicole Fleetwood, "Failing Narratives, Initiating Technologies: Hurricane Katrina and the Production of a Weather Media Event," *American Quarterly* 58, no. 3 (2006): 767–89, and for a general narrative of the event, Douglas Brinkley, *The Great Deluge: Hurricane Katrina, New Orleans, and the Mississippi Gulf Coast* (New York: Morrow, 2006); on the deterioration of economic and social conditions in postwar years, see Kent B. Germany, *New Orleans after the Promises: Poverty, Citizenship, and the Search for the Great Society* (Athens: University of Georgia Press, 2007), and Juliette Landphair, "'The Forgotten People of New Orleans': Community, Vulnerability, and the Lower Ninth Ward," *Journal of American History* 94 (December 2007): 837–45; on the Mardi Gras industry and its impact on the city's social fabric, see Kevin Fox Gothman, *Authentic New Orleans: Tourism, Culture, and Race in the Big Easy* (New York: New York University Press, 2007).

4. Bradford Gray and Kathy Hebert, "Hospitals in Hurricane Katrina: Challenges Facing Custodial Institutions in a Disaster," *Journal of Health Care for the Poor and Underserved* 18, no. 2 (2007): 283–98.

5. See Susanna M. Hoffman and Anthony Oliver-Smith, eds., *Catastrophe and Culture: The Anthropology of Disaster* (Santa Fe, N.M.: School of American Research Press, 2002); Kevin Rozario, *The Culture of Calamity: Disaster and the Making of Modern America* (Chicago: University of Chicago Press, 2007); Ted Steinberg, *Acts of God: The Unnatural History of Natural Disaster in America* (New York: Oxford University Press, 2000).

6. The expression is culled from Romain Huret, *Katrina, 2005: L'Ouragan, l'Etat, et les Pauvres* (Paris: École des Hautes Études en Sciences Sociales, 2010). See also Van Heerden Ivor and Bryan Mark, *The Storm: What Went Wrong and Why during Hurricane Katrina—The Inside Story from One Louisiana Scientist* (New York: Viking, 2006), and Walter Brasch, *"Unacceptable": The Federal Response to Hurricane Katrina* (Charleston: BookSurge, 2005).

7. George W. Bush, *Decision Points* (New York: Crown, 2010).

8. Robert Tracinski, "An Unnatural Disaster : A Hurricane Exposes the Man-Made Disaster of the Welfare State," *Intellectual Activist,* www.intellectualactivist.com/php-bin/news/showArticle .php?id=1026; Sara Le Menestrel and Jacques Henry, "La Figure du *Survivor*: Usages de la Mémoire et Gestion de la Catastrophe en Louisiane après les Ouragans Katrina et Rita," *Ethnologie Française* 40, no. 3 (2010): 495–508; Eduardo Bonilla-Silva, *Racism without Racists: Color-Blind Racism and the Persistence of Racial Inequality in the United States* (Lanham, Md.: Rowman and Littlefield, 2003).

9. See Edward T. Linenthal, *The Unfinished Bombing: Oklahoma City in American Memory* (Oxford, U.K.: Oxford University Press, 2001).

"Two Centuries of Paradox"

The Geography of New Orleans's African American Population, from Antebellum to Postdiluvian Times

RICHARD CAMPANELLA

The residential settlement patterns of New Orleans's African American population reflect, among other drivers, the city's centuries-old Franco-Hispanic-Afro-Caribbean heritage; urban slavery and emancipation; *de jure* and *de facto* racial discrimination; the geography of amenities, nuisances, and hazards; experimentation in public housing; poverty and social vulnerability; and catastrophes such as the Hurricane Katrina–induced deluge. The modern city, as a result, exhibits African American geographies that are spatially racially segregated in many ways, yet still more integrated than many American cities. Paradoxically, however, while black and white New Orleanians are now more likely to work, dine, and shop in integrated fashions, they are less likely to live together than in the eras of segregation and slavery. These complex spatial distributions from antebellum (1803–65) to postdiluvian (2005–present) times shed new light on race-related topographical and longitudinal patterns vis-à-vis the Katrina flood.

Origins

Two main waves of kidnapped Africans arrived in New Orleans during colonial times, the first under French rule in the 1720s and a larger one under Spanish dominion in the 1780s. Coupled with the New World slave trade and natural increases, New Orleans claimed an African-ancestry population of 4,108 (compared to 3,948 whites) by the time of the Louisiana Purchase (1803). Unlike most North American cities, New Orleans retained the Caribbean notion of a "gradient" between free white and enslaved black, manifested by the somewhat privileged mixed-race middle caste known as the *gens de couleur libres* (free people of color, abbreviated in contemporary documents as "FPC"). Slaves outnumbered FPCs by a 2.1-to-1.0 ratio in the newly Americanized city, a ratio that would later equalize. Despite increasingly oppressive laws at the local and state level, more FPCs would call New Orleans home than any other southern city (and occasionally more than any American city, in both relative and absolute terms) throughout most of the antebellum era.[1] Their presence helped distinguish New Orleans and Louisiana society from the national norm. "It is worthy of remark," observed a *New York Times* journalist, "that . . . free colored persons should be so differently regarded in Louisiana. . . . [They have] acquired a *status* and influence unknown in any other city, even in the Free States. . . . [O]ne in eleven [in New Orleans works as] clerks, doctors, druggists, lawyers, merchants, ministers, printers and teachers. . . . [T]he free colored population of New-Orleans are acquiring an assimilation to the whites in education and influence . . . superior to that of any other State or city."[2]

Additional cultural diversity developed among New Orleanians of African ancestry as a result of the U.S. prohibition on international slave trading in 1808. The law shifted the movement of enslaved peoples into the hands of domestic traders (as well as international smugglers), who sent "surplus" slaves from the Upper South into the Deep South to satisfy the plantation economy's insatiable demand for field labor. More than 750,000 people were forcibly shipped from Maryland, Virginia, and adjacent states into the lower Mississippi Valley over the next fifty years, a shift in the geography of African Americans so significant that one historian described it as the "Second Middle Passage."[3] The same geographical advantages that made New Orleans the South's premier port also made it the nation's busiest slave marketplace. Journal accounts provide some idea of its size. Wrote one visitor, "There were about 1000 slaves for sale at New Orleans while I was there" in March 1830.[4] "I cannot say as to the number of negroes in the [New Orleans] market," wrote a trader in 1834, "though am of the opinion there is 12–1500 and upwards, and small lots constantly coming in." Other eyewitnesses estimated three thousand slaves for sale at a particular moment later in the antebellum era, equating to roughly one marketed slave for every five resident slaves in the city.[5] Its trafficking of human beings, wrote one historian, "had a peculiar dash: it rejoiced in its display and prosperity; it felt unashamed, almost proud."[6] Disproportionately, the victims of this domestic slave trade landed in New Orleans, making the region's African-ancestry population increasingly American in ethnicity, English in tongue, and Protestant in religion, vis-à-vis the local Francophone Catholic black Creole population with colonial-era roots. Descendents of both subgroups, which may be termed Anglo-African Americans and Franco-African Americans, retain elements of their historical ethnic distinction, and trace differing uptown-versus-downtown residential settlement patterns to this day.

Antebellum Geographies of the Enslaved

Where did African American New Orleanians reside in the antebellum era? Urban slaves who labored as domestics usually dwelled in the distinctive slant-roof quarters appended behind townhouses and cottages. Others, ranging from skilled craftsmen and artisans to hired-out laborers, lived in detached group quarters on adjacent lots and alleys. An 1817 city ordinance prohibited slaves from living "in any house, out-house, building, or enclosure" not owned by their master or representative (except with documented permission), else the slave face jail time and twenty lashes, and the master a five-dollar fine.[7] Similar laws persisted throughout the antebellum period; an 1857 city ordinance reads similarly to its predecessor from forty years prior: "That it shall not be lawful for any slave to lodge or sleep in any house or premises other than that of his owner or master," with fines and lashes awaiting white and black lawbreakers respectively.[8] The following 1831 real estate advertisement (emphasis added) exemplifies the residential adjacency arranged by masters for slaves: "To Let, a good brick house, No. 113 Casa Calvo [Royal Street], faubourg Marigny, consisting of 4 rooms, 2 closets and gallery, a kitchen, stable, coach house, and 2 wells; *also a large frame house on the adjoining lot, calculated to lodge 200 negroes.*"[9]

Most masters needed no legislative prodding to settle their slaves adjacently: spatial proximity enabled monitoring of movement and promptness of service. This so-called "back-alley" settlement pattern imparted an ironic spatial integration into the city's antebellum racial geography, despite the severe and oppressive social segregation of chattel slavery. Master and slave, white and black, lived steps away from each other. Dempsey Jordan, born a slave in New Orleans in 1836, described just that arrangement when interviewed a century later: "Our quarters was small, one room house *built in the back yard of Maser's home* . . . out of rough lumber like [a] smoke house."[10] Not unique to New Orleans, the intermixed back-alley pattern has been documented in Charleston, Washington, and Baltimore.[11]

Antebellum Geographies of the Free People of Color

FPCs, who unlike their enslaved brethren *chose* their residences, rarely settled in the new "uptown" faubourgs upriver from the urban core. Rather, they clustered in the lower French Quarter (Old City), Bayou Road, the downriver faubourgs of Tremé, Marigny, New Marigny, Franklin, and those comprising the present-day neighborhood of Bywater. Evidence for this uptown-avoidance/downriver-clustering pattern comes from the 1860 census, in which FPCs equaled, and then numerically overwhelmed, the slave population as one went deeper into the lower side of town (Ward Four through Ward Nine). But on the upper side of town, the opposite occurred: in wards Three, Two, One, Ten, and Eleven, slaves consistently and greatly outnumbered free people of color. The uptown First Ward, for example, was home to seven black slaves for every FPC, while in the downtown Eighth and Ninth wards could be found three FPCs for every black slave.[12] Why did free people of color prefer the lower city? Culturally, this was the Francophone, Catholic, locally descended (Creole) side of town, an environment that most FPCs found more conducive to live, work, raise their families, and prosper. The English-speaking world on the upper side of town was not only culturally foreign terrain, but its predominantly Anglo inhabitants were more hostile to the very notion of free people of color as a legally recognized caste. The regions of the interior and upper South from which many Anglos migrated generally maintained a "rigid, two-tiered" racial caste system.[13]

In addition to the Creole cultural driver, the flow direction of the Mississippi River and the topography of New Orleans played roles in spatially sorting (and reinforcing) FPC residential settlement patterns. Hydrologically, the side of town that was culturally Creole was also juxtapositioned downriver from the urban core—consequential, because the city's sewerage and detritus, which by law had to be "emptied into the current of the river,"[14] polluted the water source and environs of lower-city denizens. Being downriver tainted the residential desirability of those neighborhoods, which lowered their property values and motivated the construction of humble abodes. Cheap real estate repelled prospective home buyers who were wealthier and white, and attracted those who were poorer and more likely to be black (that is, FPC). By their very presence, FPCs drew more members of their intermediary caste to settle nearby, via social and familial networks based on ethnic commonalities. Thus formed a spatial racial pattern of black settlement in the downriver eastern half of the metropolis which would prove enormously consequential centuries later.

Another physical driver of FPC settlement entailed New Orleans's depositional geology. A river in a deltaic environment builds land by periodically overtopping its banks and depositing alluvium laterally. The lion's share of the coarsest sediment settles closest to the channel, while smaller amounts of finer-grain particles come to rest farther away. Over millennia, this process raises lands nearest the river highest, while areas increasingly distant from the river taper downwardly—precisely the opposite of far-more-common erosional landforms, in which water erodes and lands closest to rivers lie lowest. Human settlement since prehistoric times, and particularly since the founding of New Orleans in 1718, exploited the higher, drier riverine elevations (natural levees) and avoided the low-lying flood-prone mosquito-infested backswamp. Just as living upriver evaded environmental nuisances and living downriver increased them, living on higher ground reduced environment hazards while living near the swamp augmented them. Empowered members of the white caste (plus their slaves) thus monopolized the high ground—dubbed the "front of town" in local parlance—while free people of color gravitated to the "back of town."

A Case Study

Evidence for the front-/back-of-town racial geography comes from a spatial analysis of the 1820 census, focusing on the six river-parallel streets of the French Quarter: Levee (now Decatur), Chartres, Royal, Bourbon, Dauphine, and Burgundy. Because they share the same developmental era, size, geometry, culture, urban granularity, and population density, these streets present a fine comparative cross-section of antebellum New Orleans. Areas near the Mississippi, in the front of town along the Levee Street corridor, hosted most of the city's riverfront economic activity. Beneath them lay the highest, best-drained, and most arable soils of the entire deltaic plain. Thus, commercial activity and prosperity generally gravitated toward the front of town—but not *too* close, because the river's literal edge abounded with bustling traffic, malodorous and cacophonous port nuisances, and unsavory characters. This explains why the Levee Street corridor, now Decatur and North Peters, is replete to this day with commercially oriented storehouses, and nearly devoid of residences.

Areas far from the river, in the back of town along and behind Burgundy Street, saw none of those riverfront annoyances. Instead, however, they suffered from low topographic elevation and proximity to the flood-prone mosquito-breeding swamps. Medical theories of the day held that vapors rising from wetlands (miasmas) caused diseases such as yellow fever, and while miasmatic theory later proved to be apocryphal, the perception of risk created its own reality and further undermined the area's socio-economic valuation. Far away from the economic heartbeat of the city, the low-rent back of town attracted the working class, the poor, and the disenfranchised. This explains why Burgundy Street (and neighborhoods behind it) abounds to this day with simple cottages and shotgun houses, with relatively few storehouses, townhouses, or mansions.

Areas *between* the riverfront and the back of town afforded the benefits of the convenient heart of the city, close to the economic activity, and on relatively higher ground, while maximizing the distance from both riverfront nuisances and backswamp hazards. That spatial positioning translated to higher real estate values, better housing, and thus a wealthier, more-white demographic that was more likely to comprise members of the master class. This explains why Chartres and Royal streets had the highest percentage of slaves in 1820, with 37.3 and 34.8 percent respectively. It explains why John Adems Paxton observed in the 1820s that on "the streets nearest the river, the houses are principally of brick . . . but in the back part of the town, they are generally of wood."[15] It also explains why, to this day, opulent townhouses and mansions proliferate on Chartres and Royal but remain scarce on Burgundy. The intermediary Bourbon and Dauphine streets represent increments between those more-desirable and less-desirable spaces, hence their appeal to the white working-to-middle class. Those two streets were 30 percent enslaved, markedly less than Chartres and Royal but more than the back of town.

We see these human geographies manifested with intriguing linearity in the 1820 census. For example, each street's percentage of white residents decreased steadily with increasing distance from the river. Levee Street, closest to the river, was 60.4 percent white; next, Chartres Street, was 56.9 percent; Royal 56.6 percent; Bourbon 49.5 percent; Dauphine 39.5 percent; and back-of-town Burgundy at only 33.4 percent. Why? Whites, occupying the most privileged position in New Orleans's three-caste social structure, had more fiscal and legal wherewithal to position themselves closer to amenities, conveniences, and commerce, and farther from nuisance and risk. Bourbon's middling spatial position put it almost precisely in a middling racial-composition position (49.5 percent white), nearly matching the city's 46-percent-white overall population.[16]

Similarly linear patterns arise when we map out the percentage of FPC per street. Back-of-town Burgundy Street had the highest FPC percentage among our six streets, at 37.6 percent. That percentage then steadily decreased as we move closer to the river, higher in elevation, and upward in real estate value: Dauphine had 29.3 percent; Bourbon 21.7 percent; Royal 8.6 percent; and Chartres and Levee only 5.8 and 5.6 percent, respectively. Comparing this FPC pattern to that of slaves, we may surmise that African Americans, when enslaved, had little choice but to live in or near the privileged space of the white master households, whereas when they were free, they found themselves relegated to the inferior space of the lower neighborhoods and the back of town.

Another fascinating front-/back-of-town human geography involved gender. Men and women did not evenly distribute themselves across the streetscape; rather, males numerically predominated in the front of town while females made up nearly two-thirds of humanity in the back of town. Levee Street was 46.1 percent female, Chartres 46.6 percent, Royal 54.9 percent, Bourbon 58.4 percent, Dauphine 63.2 percent, and Burgundy 65.1 percent. The male predomination of the front can be explained by the large number of single men or short-term residents living alone who participated in port-related riverfront commerce. The female predomination of the back of town, which correlates with this area's high FPC population, reflects the low economic status of female-led households, including women living alone.[17]

The geography of black New Orleans, then, consisted of slaves intricately intermixed citywide and FPCs predominating in the older, lower neighborhoods. A journalist articulated these patterns in an 1843 *Daily Picayune* article in the racialized lexicon of the day: "The Negroes are scattered through the city promiscuously; those of mixed blood, such as Griffes, Quarteroons, &c., [free people of color] showing a preference for the back streets of the First [French Quarter, Faubourg Tremé] and part of the Third Municipality [Faubourg Marigny and adjacent areas]."[18]

Postbellum Geographies and the Beginnings
of Racial Spatial Disassociation

New Orleans's black population surged by 110 percent between the censuses of 1860 and 1870, bracketing the trauma of Civil War and the ensuing social and political tumult. It rose another 54 percent by the turn of the century.[19] Behind the numbers were thousands of emancipated people abandoning agrarian toil for the hope of the metropolis. Caught up in its own woes, the unwelcoming city of New Orleans nevertheless offered better opportunities to freedmen than the cane and cotton fields. In 1870, black men, who made up one-quarter of the labor force, worked 52 percent of New Orleans's unskilled labor jobs, 57 percent of the servant positions, and 30 to 65 percent of certain skilled positions.[20]

Where were these rural-to-urban immigrants to reside? Unaffordable housing and racially antagonistic neighbors prevented the freedmen from settling in most front-of-town areas in the high-density urban core on the high ground near the river. Townhouses in the inner city, many recently vacated by wealthy families who lost their fortunes or otherwise moved to new streetcar suburbs, had since been subdivided into low-rent apartments, but these hovels were more likely to be leased to poor immigrants than to poor rural blacks. Nor could the freedmen easily take refuge in the lower faubourgs of the former free people of color, who often scorned the poverty-stricken newcomers as threats to their once relatively privileged (but now rapidly diminishing) social status.

Destitute and excluded, most emancipated families had little choice but to settle in the ragged back of town, where urban development dwindled into amorphous low-density shantytowns and eventually dissipated into low-lying deforested swamps. The back of town offered low real estate costs because its environmental hazards, urban nuisances, inconveniences, and lack of amenities and city services rendered it the land of last resort. Lowly shotgun houses and common-wall cottages arose to accommodate the demand for low-income housing, thus reinforcing the poverty. The freedmen joined other socially marginalized people already settled at the *physically* marginalized geography that was the backswamp, in the formation of the city's first large-scale, exclusively black neighborhoods. Among the areas that coalesced with black populations (predominantly Anglo-African American) in the postbellum era was the present-day neighborhood of Central City, which, to a remarkable degree, limns a racial geography that corresponds to edge of the former backswamp.

Concurrently, emancipation rendered obsolete the laws requiring adjacent master-slave housing; white families in the wake of the Confederate defeat now winced at the notion of their former slaves living next door. Indeed, the so-called back-alley pattern of racial proximity had started to wane even before the war as Irish and German immigrants, arriving in the 1850s, flooded the labor market and began to replace enslaved domestics. In the postbellum era, then, New Orleans's back of town grew increasingly black in both absolute and relative numbers, while the front of town became more white: initial evidence of the racial spatial disassociation that would define the human geography of New Orleans for a century to come.

Persisting Spatial Heterogeneity amid Increasing
Spatial Homogenization

Yet complicating patterns of racial spatial heterogeneity persisted from earlier times. Creoles of color (a group generally synonymous with FPCs before
the war, plus their descendants) continued to choose their neighborhoods on
their terms, for reasons of tradition, family, religion, culture, convenience,
economics, or real estate. They usually remained on the lower side of the
city. Other black families, whose men of the house worked on the docks and
wharves, settled near the riverfront for its proximity to the port. Still others resided in areas that, unlike the low-lying back of town, lay counterin-
tuitively high on the natural levee and free from flood threat—but whose urban nuisances nevertheless rendered them less desirable and lower in rent.
These front-of-town areas included blocks immediately along the riverfront
wharves, docks, railroads, warehouses, and cotton presses: noisy, smelly, bustling zones teeming with workers tending to ships arriving round-the-clock,
places that wealthy families disdained. To this day, Tchoupitoulas Street, the
highest-elevation street in the city and the safest from Katrina-style flooding,
nevertheless is home to a largely working-class black population. Other areas
of high urban nuisance and hazard through the city, such as those blocks near
factory and industrial sites, municipal dumps, cemeteries, hospitals, canals,
and railroad tracks, also attracted the development of cheap housing stock,
to which gravitated—or were relegated, given their lack of other options—the
disproportionately African American underclass.

Still other black families settled in what has been described as the "su-
perblock" pattern. Because many African Americans worked as domestics
for wealthy uptown whites, they (together with working-class whites) often
settled in small cottages and shotgun houses developed in the "nucleus" of
the "superblocks"[21] outlined by the great mansion-lined avenues such as
St. Charles, Louisiana, Napoleon, Nashville, State, and Carrollton. Those
grand avenues were developed for upper-class residential living because of
their spaciousness, magnificence, see-and-be-seen perches, and proximity
to streetcar service. Narrower streets within the nucleus of the avenue grid,
on the other hand, were built up with cheaper, humbler housing stock. The
grand avenues thus formed a lattice of upper-class whites around cores of
working-class blacks and whites, many of whom could conveniently walk to
their domestic jobs in the mansions. This phenomenon imparted a certain in-
terracial proximity that survives to this day throughout uptown New Orleans.

Thus, even as the city's relatively intermixed racial geographies gradually disassociated (that is, pulled apart and began forming more homogenous clusters) after the Civil War, they remained far more spatially heterogeneous than those of northern cities. The German geographer Friedrich Ratzel noticed the pattern a decade after emancipation and offered three hypotheses: "New Orleans has a larger colored population than Charleston or Richmond, but you would not believe it if the statistics did not say so—*so much less is the distance separating these people from the whites.* This is partly because of the great preponderance of mulattoes (who call themselves "yellow"... as opposed to "black"...), partly because of prosperity that prevails in these circles, and partly, though not least of all, *because the French in Louisiana never set themselves off so strictly from their slaves and freed men as the Anglo-Americans did in the other slave states.*"[22]

Twentieth-Century Geographies and the Acceleration of Racial Spatial Disassociation

Two national trends around the turn of the twentieth century further pulled apart New Orleans's historically intermixed racial geography. One commenced—or rather climaxed—with *Plessy v. Ferguson* (163 U.S. 537) in 1896. That landmark U.S. Supreme Court decision (on a New Orleans–based case) to legalize "separate but equal" statutes marked the culmination of decades of increasingly polarizing racial tension in the wake of Emancipation. It also represented a final act in the century-long process of transforming New Orleans's old Franco-Caribbean recognition of a gradient between black and white (via a legally institutionalized intermediary FPC caste) into an American style "rigid, two-tiered [social] structure that drew a single unyielding line between the white and nonwhite."[23] After *Plessy,* legally sanctioned racial segregation would affect real estate sales and deed covenants (and therefore future residential settlement patterns), as well as access to public schools, jobs, public housing, recreational facilities, retailers, food services, and nearly every other aspect of life in New Orleans and the South.

The second trend entailed Progressive Era municipal reforms, which, in New Orleans and elsewhere, brought significant improvements to potable water, sewerage, public health, electrification, transportation, and most importantly for this deltaic city, drainage and flood control. The installation of technologically advanced municipal drainage apparatus during 1895–1915 radically reworked the physical geography of the city, draining groundwater from the backswamp at the same time that federally engineered riverfront levees largely ended the threat of river floods, and new lakefront levees reduced flood risk from Lake Pontchartrain. These thrilling new advances, in the minds of most New Orleanians, seemingly neutralized the backswamp's low waterlogged soils as an environmental risk, and allowed modern development to extend upon them. "The entire institutional structure of the city was complicit" in the ensuing urbanization of the lowlands, wrote one local historian; "developers promoted expansion, newspapers heralded it, the City Planning Commission encouraged it, the city built streetcars to service it, [and] the banks and insurance companies underwrote the financing."[24] Automobiles arrived serendipitously, followed by modern transportation arteries. Developers eagerly built new subdivisions—Lakeview and Gentilly, for example—in the spacious, modern, automobile-conducive California style, quite the antithesis of the antique housing stock that predominated in the rest of the city. They also installed racist deed and housing covenants explicitly prohibiting sale or rental to black families.

The new subdivisions were a hit. During the 1910s–1940s, middle-class white families, formerly residents of the historical front of town, leapfrogged over the black back of town and settled in the low-lying, whites-only lakeside subdivisions. Between 1920 and 1930, nearly every census tract lakeside of the Metairie/Gentilly Ridge *at least* doubled in population. Low-lying Lakeview saw its population increase by about 350 percent, while parts of equally low Gentilly grew by 636 percent. Older neighborhoods on higher ground, meanwhile, lost residents: historic faubourgs Tremé and Marigny dropped by 10 to 15 percent; the French Quarter declined by one-quarter. The high-elevation Lee Circle area lost 43 percent of its residents, while low-elevation Gert Town increased by a whopping 1,512 percent.[25] Similar figures could be cited for the 1910s and 1930s–50s (see figs. 1.1–1.7).

The lakeside urbanization provided geographical space at the right historical moment for *de jure* racism to manifest itself in the emerging residential settlement patterns, and modern urban planning played a key role in the coordination. This era (1910s–20s) saw the emergence of city planning commissions and zoning ordinances throughout American cities, aimed at protecting homeowners' property values by segregating residential, commercial, and industrial land uses. Southern city planners viewed black neighborhoods as affecting property values, and proceeded to develop methods of segregating them. That the U.S. Supreme Court struck down Louisville's racial zoning ordinance in the 1917 case of *Buchanan v. Warley* (245 U.S. 60) did little to dissuade southern urban planners from such blatantly unconstitutional policies. A year after New Orleans created its own City Planning and Zoning Commission in 1923, the Louisiana legislature passed a law to segregate racially the state's larger cities. New Orleans promptly complied by adopting a municipal code in 1924 forbidding building permits for black-occupied buildings in white neighborhoods (and vice-versa), and preventing rentals to blacks without the support of white neighbors. When the U.S. Supreme Court again struck down that law in *Harmon v. Tyler* (273 U.S. 668), New Orleans authorities resorted to two trusty methods to keep the races apart: racially discriminatory deed and housing covenants, which had been legally upheld in state courts, and the supportive complicity of the real estate industry. A 1924 article entitled "Segregation by Co-operation of Civc [*sic*] Bodies" cheerfully reported how a committee of housing-market players stepped in to do the work of Jim Crow after courts struck down legislative mechanisms. "There is considerable present interest in sectional segregation," declared the *Times-Picayune,* "which many consider important [for] residential development and values." Toward this end, real estate agents J. A. and W. G. Moran formed a committee whose

> primary province will be to specifically denote residential areas for whites and colored. Such lines established, the next step will be for individuals and associations who perform any of the functions incidental to ownership . . . to pledge not to participate in any transaction in which either white or colored would attempt to obtain residence in any section reserved for the opposite race. The real estate man would decline to sell or lease, the lawyer to examine title, the notary to pass the act, the insurance man to protect, the architect to design or remodel, and the homesteads to grant loans, where any such invasion would be intended. Co-operation along those lines would be more effective than formal law. . . . The outcome would guarantee that both races in New Orleans would continue to reside here in peace and tranquility.[26]

Through *de jure* tools such as racist deed covenants, civic noncooperation in real estate transactions, and the *de facto* racism of many citizens, the intricately intermixed racial geographies of the nineteenth century spatially disassociated into the white and black neighborhoods of the twentieth century. Strictly segregated public housing further entrenched the trend, as did public schools.

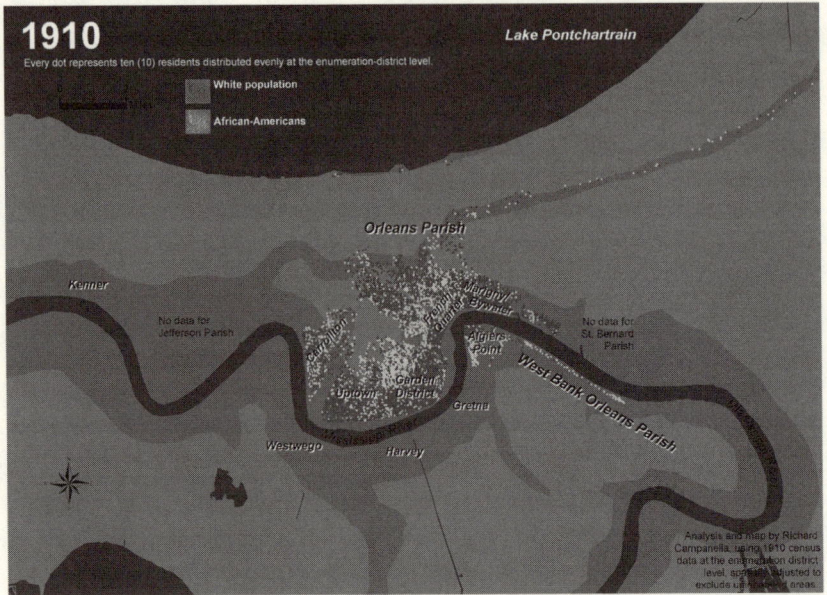

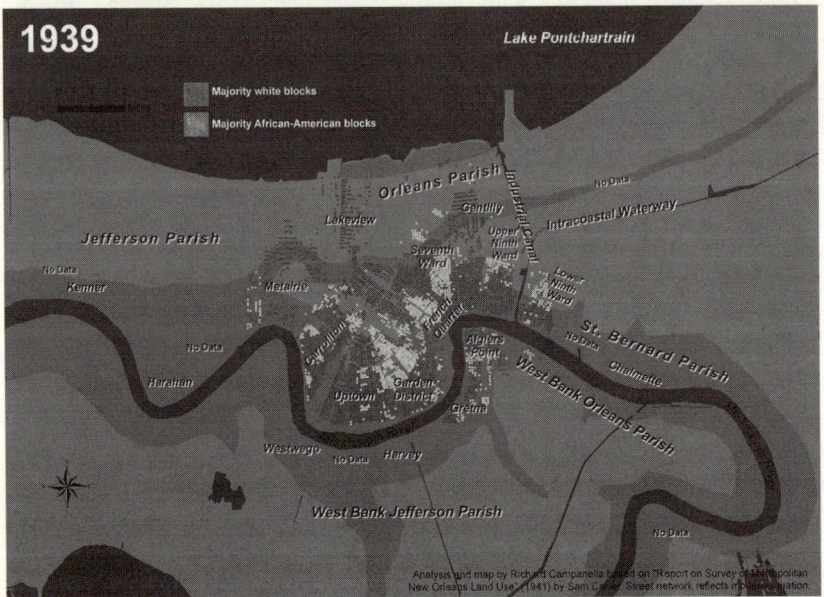

Figs. 1.1–1.4. Residential settlement patterns by race, New Orleans, 1910–1960. Maps by author.

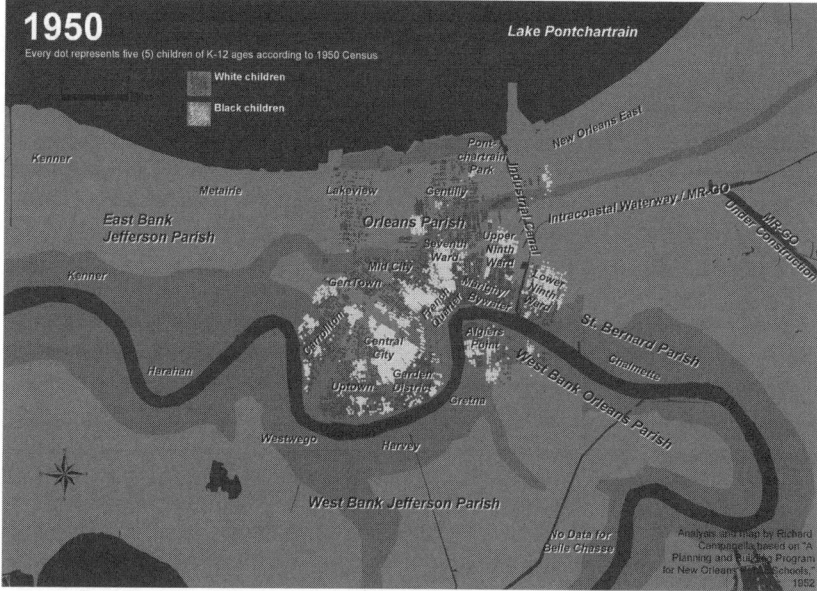

1950
Every dot represents five (5) children of K-12 ages according to 1950 Census

White children

Black children

Lake Pontchartrain

New Orleans East

Pontchartrain Park

Kenner

Metairie

Lakeview

Gentilly

Industrial Canal

Intracoastal Waterway / MR-GO

Under Construction

MR-GO

East Bank Jefferson Parish

Orleans Parish

Seventh Ward

Upper Ninth Ward

Kenner

Mid City

Gert Town

Marigny Bywater

Lower Ninth Ward

Carrollton

Central City

Algiers Point

St. Bernard Parish

Chalmette

Harahan

Uptown

Garden District

West Bank Orleans Parish

Gretna

Westwego

Harvey

West Bank Jefferson Parish

No Data for Belle Chasse

Analysis and map by Richard Campanella based on "A Planning and Building Program for New Orleans Public Schools," 1952

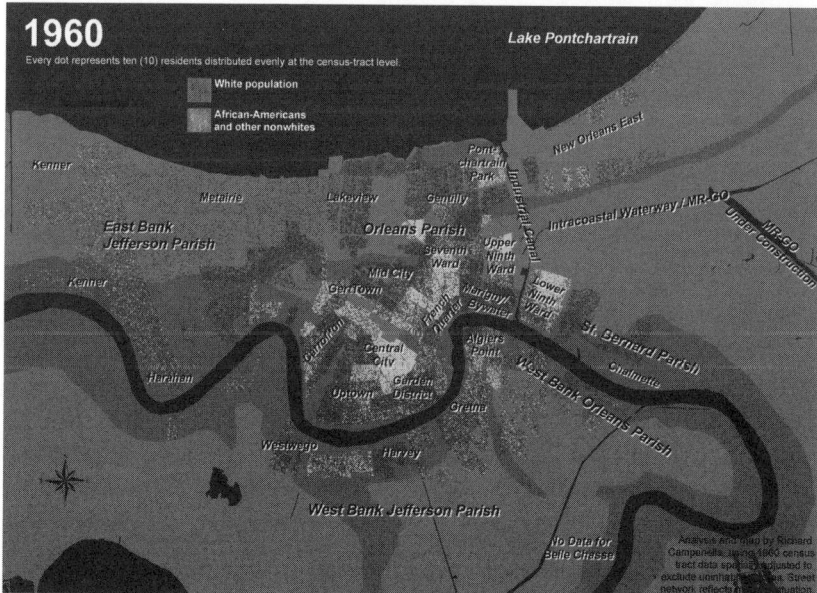

1960
Every dot represents ten (10) residents distributed evenly at the census-tract level.

White population

African-Americans and other nonwhites

Lake Pontchartrain

New Orleans East

Pontchartrain Park

Kenner

Metairie

Lakeview

Gentilly

Industrial Canal

Intracoastal Waterway / MR-GO

Under Construction

MR-GO

East Bank Jefferson Parish

Orleans Parish

Seventh Ward

Upper Ninth Ward

Kenner

Mid City

Gert Town

Marigny Bywater

Lower Ninth Ward

Carrollton

Central City

Algiers Point

St. Bernard Parish

Chalmette

Harahan

Uptown

Garden District

West Bank Orleans Parish

Gretna

Westwego

Harvey

West Bank Jefferson Parish

No Data for Belle Chasse

Analysis and map by Richard Campanella, using 1960 census tract data spatially adjusted to exclude uninhabited areas. Street network reflects that of the era.

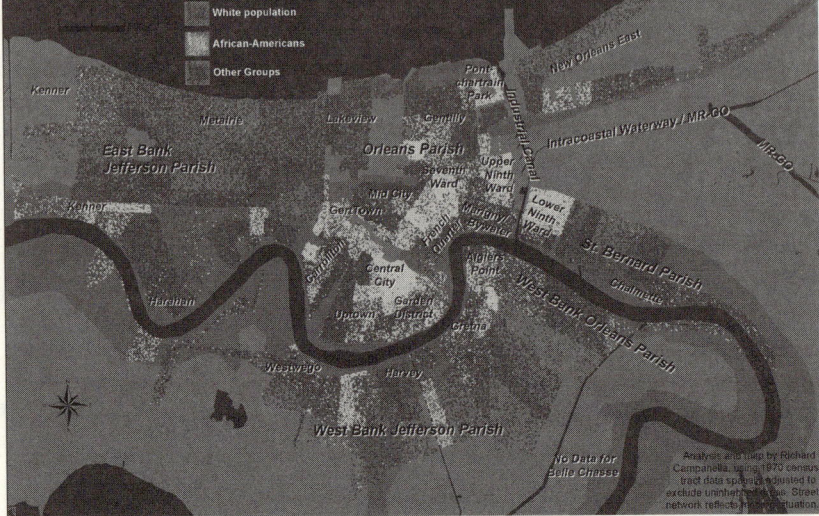

1970
Every dot represents ten (10) residents distributed evenly at the census-tract level.

White population
African-Americans
Other Groups

Lake Pontchartrain

Kenner

Metairie

Lakeview

Gentilly

New Orleans East

Pontchartrain Park

Industrial Canal

Intracoastal Waterway / MR-GO

MR-GO

East Bank
Jefferson Parish

Orleans Parish

Seventh
Ward

Upper
Ninth
Ward

Kenner

Mid City

Gert Town

French
Quarter

Marigny
Bywater

Lower
Ninth
Ward

St. Bernard Parish

Chalmette

Gretna

Algiers
Point

West Bank Orleans Parish

Central
City

Garden
District

Harahan

Uptown

Westwego

Harvey

West Bank Jefferson Parish

No Data for
Belle Chasse

Analysis and map by Richard
Campanella, using 1970 census
tract data spatially adjusted to
exclude uninhabited area. Street
network reflects recent situation.

2000
Every dot represents ten (10) residents distributed evenly at the census-block level.

White population
African-American
Asian ancestry
Hispanic ethnicity

Lake Pontchartrain

Kenner

Metairie

Lakeview

Gentilly

New Orleans East

Versailles

Pontchartrain Park

Industrial Canal

Intracoastal Waterway / MR-GO

East Bank
Jefferson Parish

Orleans Parish

Seventh
Ward

Upper
Ninth
Ward

Kenner

Mid City

GertTown

French
Quarter

Marigny
Bywater

Lower
Ninth
Ward

St. Bernard Parish

Chalmette

Gretna

Algiers
Point

West Bank Orleans Parish

Central
City

Garden
District

Harahan

Uptown

Westwego

Harvey

West Bank Jefferson Parish

Belle
Chasse

Analysis and map by
Richard Campanella,
using 2000 census
tract/block level.

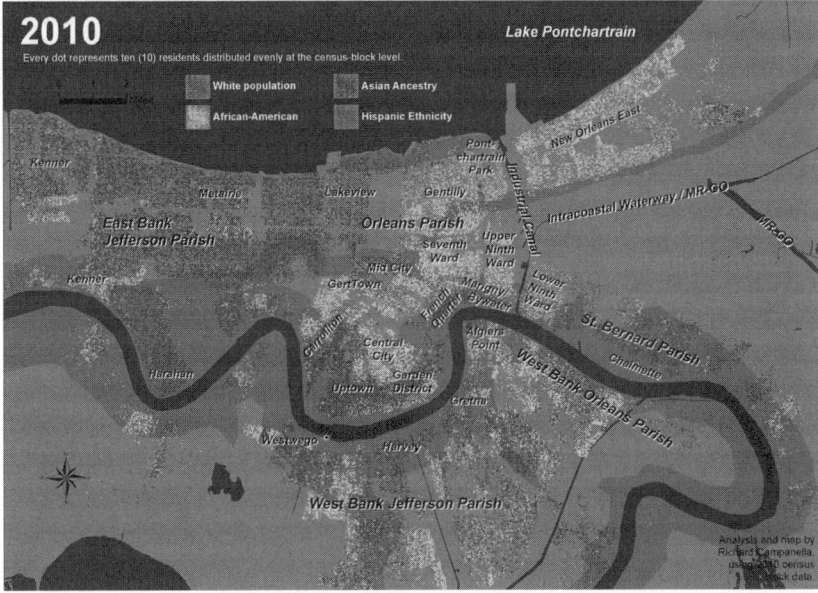

2010

Every dot represents ten (10) residents distributed evenly at the census-block level

Lake Pontchartrain

White population   Asian Ancestry

African-American   Hispanic Ethnicity

Kenner

Metairie   Lakeview   Gentilly

Pontchartrain Park

New Orleans East

East Bank Jefferson Parish

Orleans Parish

Industrial Canal

Intracoastal Waterway / MR-GO

MR-GO

Kenner

Seventh Ward   Upper Ninth Ward

Mid City

Gert Town

French Quarter   Marigny / Bywater   Lower Ninth Ward

St. Bernard Parish

Central City

Algiers Point

West Bank Orleans Parish

Chalmette

Harahan

Garden District   Uptown

Gretna

West Bank Orleans Parish

Westwego   Harvey

West Bank Jefferson Parish

Analysis and map by Richard Campanella, using 2010 census block data.

Figs. 1.5–1.7. Residential settlement patterns by race, New Orleans, 1970–2010. Maps by author.

A 1952 study by the Orleans Parish School Board's Office of Planning and Construction illustrates the key role that sophisticated empirical urban planning played in managing the new geography of race. Following helpfully expository opening chapters with headings like "Why Plan" and "What to Plan," the report carefully enumerated and mapped "white" versus "negro" children at the block level, predicted the future fecundity of their parents, estimated the accuracy of the predictions, and proposed a suite of similarly segregated new education facilities, thus ensuring another generation of residential segregation.[27] By the mid-twentieth century, white and black New Orleanians were moving away from each other *en masse,* and the trend would only strengthen.

Tremendous social transformations forged new racial relationships in the latter half of the century. Chief among these were *Brown v. Board of Education of Topeka* (347 U.S. 48, 1954), the Civil Rights Act of 1964, and the ensuing desegregation of public facilities, integration of public schools, and increased opportunities in education, employment, and housing for African Americans. Jim Crow disappeared with less violence and resistance here than in other southern cities; black and white New Orleanians subsequently found themselves working, shopping, and dining together in increasing numbers (fig. 1.8). Yet *living* together did not necessarily follow; in fact, residential integration diminished. Suburban-style subdivisions in Jefferson, St. Bernard, and St. Tammany parishes, even as far as coastal Mississippi, drew white New Orleanians by the tens of thousands between the censuses of 1960 and 2000. Middle-class uptown whites generally gravitated westward into Metairie and Kenner suburbs of Jefferson Parish; working-class downtown whites usually resettled eastward in St. Bernard Parish or on the West Bank.

Middle-class African Americans, for their part, mostly moved lakeward to the neighborhoods east of City Park and thence into the subdivisions of eastern New Orleans. This longitudinal diffusion, which has garnered little scholarly attention but is important in understanding modern racial geographies and Katrina impacts, can be explained by a number of reasons. First, many middle-class black New Orleanians descended from the Creole of color / FPC population, who lived on the lower (eastern) side of the nineteenth-century city. When the eastern lakeside lowlands were drained in the early 1900s, members of this predominantly Franco-African American ethnicity often simply expanded into those new adjacent neighborhoods of the Seventh, Eighth, and Ninth wards. Other lakefront subdivisions were off-limits to black residency—until wealthy white philanthropists in the 1950s funded the first modern sub-

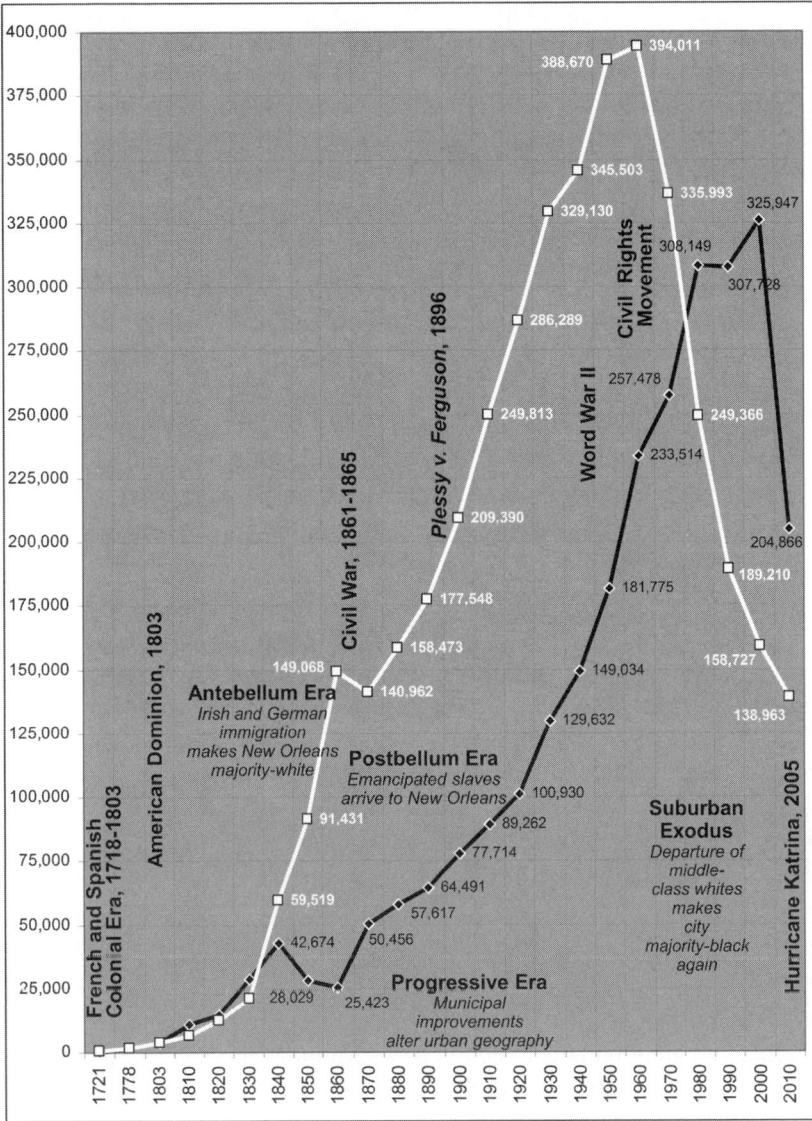

Fig. 1.8. Population of New Orleans by race, 1721–2010. White line and figures represent population of whites and other nonblacks in New Orleans (Orleans Parish) according to the census of that year. Black line and figures depict African Americans (both enslaved and free for 1721–1860). Research and graphic by author.

urban-style subdivision (complete with a golf course and curvilinear streets) for black families. Completed in 1955 and home to the black-only Southern University starting in 1959, Pontchartrain Park formed an isolated African American enclave on the lakefront, surrounded on two sides by water and the other two sides by unfriendly white neighbors. After integration in the 1960s, black families were more likely to migrate into the east-of-City-Park neighborhoods of greater Gentilly in part because Pontchartrain Park served as a precedent for middle-class black occupancy in this area. Nearby Dillard University and the St. Bernard Public Housing Development, both predominantly black, played similar roles in the otherwise all-white subdivisions east of City Park. The neighborhood *west* of City Park, Lakeview, had none of these islands of early black residency, and remains mostly white to this day. It should also be noted that there is far more developable acreage in Orleans Parish east of City Park than there is to the west; with more space come more opportunities for recent human geographies to take root.

Much of that open expanse lay east of the Industrial Canal. Known today as "New Orleans East," this twenty-square-mile ridge-intersected marsh remained mostly undeveloped into the 1960s. After Hurricane Betsy's surge flooded this and adjacent hydrological sub-basins in 1965, authorities built hurricane-protection levees around the basin and upgraded drainage canals and pumps to remove surface runoff and groundwater. In a manner similar to the Progressive Era transformations in the heart of the city, these improvements seemingly eliminated the risk of inundation and created valuable real estate. Motivated by the petroleum industry–fueled housing boom as well as the promise of well-paying jobs at the new NASA Michoud assembly plant, developers built numerous suburban-style subdivisions replete with grade-level ranch houses and split-levels. White middle-class New Orleanians subsequently moved in large numbers eastward over the Industrial Canal and into these new subdivisions in the 1970s, in the same manner that their parents and grandparents migrated lakeward into Lakeview and Gentilly in the 1910s–40s. But this did not last long: with the oil bust of the 1980s, the plateauing of NASA jobs, declining public schools, new multi-family housing with subsidized rents, and increased racial tensions in society and city government, "New Orleans East" swung dramatically from mostly white to mostly black between the 1970s and 1990s. By century's end, the greater New Orleans metropolitan area had racially dichotomized into a mostly white West and a mostly black East, with notable exceptions traceable to historical times. Ironically, it formed more segregated racial spatial patterns than it did in the early 1800s. "Two centuries of paradox" is how one researcher described the phenomenon.[28]

Public Housing Projects as Racial Spatial Concentrators

A key driver of twentieth-century racial spatial disassociation began as a progressive federal and city government program designed to help the poor. Following the U.S. Housing Act of 1937, the Housing Authority of New Orleans (HANO) cleared a number of old neighborhoods, home to thousands and replete with nineteenth-century architectural gems but considered "blighted" at the time, to make room for subsidized housing for struggling families. Two- and three-story common-wall brick apartments, tastefully designed with local architectural styles and scale, were built in geometrical arrangements among grassy walkways and oak trees. Each complex was racially segregated: two white-only developments were located on higher elevation and closer to the front of town, while the four black-only projects occupied lower spots in the back of town. The complexes expanded following the Housing Act of 1949. After desegregation of the projects in the 1960s, white occupants departed for affordable-living alternatives in working-class suburbs, and poor blacks took their places. The *de jure* segregation that had governed occupancy since the 1940s had, by the 1960s–70s, transformed to *de facto,* as tens of thousands of the city's poorest African Americans became intensely spatially consolidated into a dozen projects. Worse yet, all the housing projects were cut off from the street grid and isolated from adjacent neighborhoods. What resulted was the familiar litany of structural social problems: intergenerational dependency, under-education and unemployment, teen pregnancy, drug trading, gang activity, and incessant violent crime. (Whether the projects bred and exacerbated social ills, or merely spatially concentrated them, is a matter of ongoing debate.)

The deteriorating on-site conditions amid HANO's dysfunctional oversight motivated the intervention of the U.S. Housing and Urban Development (HUD) during the George H. W. Bush and Clinton administrations. HUD's new philosophy, encapsulated in a controversial concept named Project HOPE ("Homeownership and Opportunity for People Everywhere"), called for the demolition of troubled projects in cities nationwide, followed by their replacement with mixed-income New Urbanist communities in which subsidized rental units for the poor abutted market-rate rentals and purchasable homes aimed at modest-income families. Design guidelines involved the reestablishment of the street grid plus the creation of sidewalks, green space, minimum setback distances, individualized access between private and public space (rather than common hallways), revived historical architectural styles with an emphasis on porches and galleries, and an abundance of pastel-colored paint. The HOPE philosophy rested on two debatable notions: that a physically improved and aestheticized place creates a better society, and that the intermixing of classes in residential patterns restrains delinquency and dependency among the poor. HOPE also openly embraced neoliberal sensibilities in its appeal to private-sector real estate developers, via tax-increment financing and other marketplace motivators, to invest in public housing. In the late 1990s and early 2000s, amid vocal opposition but with the general support of nonprofits and the public, the solidly built circa-1940 units of the St. Thomas projects were demolished and rebuilt with pleasant cottages and townhouses inspired by local architectural traditions. Similar units in the Desire, Florida, and Fischer projects arose subsequently, while the Lafitte, St. Bernard, C. J. Peete, and B. W. Cooper projects awaited their turn. Opponents read bitter irony into the HOPE policy, noting that New Orleans's circa-1940 housing projects, with their modest scales, airy verandahs, and shady courtyards, seemed to embody New Urbanism decades before the term was coined. Paralleling Chicago's Cabrini-Green and Atlanta's East Lake attempts at mixed-income public housing (which really did replace ugly de-humanizing high-rises), New Orleans's component of the nation's grand neoliberal socio-spatial experiment in public housing got under way.

Then Katrina struck.

Race, Topographic Elevation, and Katrina's Floodwaters

Hurricane Katrina's levee-breaching floodwaters trapped about 100,000 people within city limits during the harrowing week of August 29 through September 4, 2005, interrupting that experiment and most urban life (fig. 1.9). As the catastrophe made headlines worldwide, observers remarked openly about the overwhelming preponderance of the African American poor among those stranded within the city. Many explained the disproportion as a product of racialized geographies—namely, spatial correlations between economic class and topographic elevation. The tragedy and the subsequent discourse on race and class in America brought New Orleans's historical racial geography to worldwide attention. The *Herald* in Glasgow, Scotland, reported that "the poorest groups have lived in the low-lying areas which have been devastated by Katrina."[29] The *New York Times* editorialized, "It is not a coincidence that

Fig. 1.9. Katrina flood depths, with darkest shades indicating eight to twelve feet of standing water four days after the storm. Map by author based on data from FEMA.

many of those hard-hit, low-lying areas have had poor and predominantly African American residents."[30] *USA Today* reported, "The low-lying wards that suffered the worst damage were mostly black neighborhoods."[31] A local activist commented, "Black people only moved [to low-lying Gentilly and eastern New Orleans] because all the good high ground had been taken."[32] Editors of the journal *World Watch* billed an article entitled "Race and the High Ground in New Orleans" with the sub-headline, "Poor and black = low, wet, and maybe dead."[33] Two spatial analyses help clarify these relationships: one to determine whether black New Orleanians were indeed more likely to reside on lower ground before Katrina, and another to determine whether black New Orleanians were more likely to be flooded by Katrina's surge.

To answer the first question, block-level Census 2000 racial and ethnic population data were intersected with two-foot contour intervals derived from LIDAR digital elevation models of Orleans Parish. The results are shown in table 1.1. While New Orleanians of all backgrounds could be found in large numbers both above and below sea level, blacks skewed more toward below-sea-level locales while whites leaned slightly toward above-sea-level areas. Roughly two feet of topographic elevation separated the races: the white population lived on average 0.48 feet above sea level, while African Americans averaged 1.57 feet below sea level. (All residents of New Orleans lived at an average of 1.0 feet below sea level, and only 38 percent of all New Orleanians lived above sea level in 2000.)

Were black New Orleanians more likely to reside on lower ground before Katrina? The answer is yes, but not overwhelmingly, and with major exceptions, such as predominantly white below-sea-level Lakeview and predominantly black above-sea-level Irish Channel. This was, and remains, a city whose population generally straddles sea level, with some racial disproportionality but not enough to validate unconditionally the simple linear white-

Table 1.1. Elevation zones according to race or ethnicity, 2000

Elevation zone	Total 2000 Population	White	Black	Hispanic	Asian
Higher than 13 feet **above** sea level	74	41	18	8	6
10 to 12 feet **above** sea level	3,136	2,389	594	128	40
8 to 10 feet **above** sea level	11,255	5,302	5,571	390	106
6 to 8 feet **above** sea level	20,105	10,028	9,379	769	178
4 to 6 feet **above** sea level	33,183	19,807	11,851	1,482	431
2 to 4 feet **above** sea level	44,451	17,452	25,325	1,639	488
0 to 2 feet **above** sea level	72,530	17,976	52,263	1,801	754
0 to 2 feet **below** sea level	83,571	15,868	64,661	2,410	1,035
2 to 4 feet **below** sea level	99,634	17,444	75,560	3,317	4,066
4 to 6 feet **below** sea level	54,884	15,080	35,808	1,522	2,688
6 to 8 feet **below** sea level	51,639	13,760	35,811	1,411	905
8 to 10 feet **below** sea level	7,377	564	6,405	131	292
10 to 12 feet **below** sea level	790	124	615	12	28
12 to 14 feet **below** sea level	2,041	114	1,862	25	31
14 to 16 feet **below** sea level	1,047	91	940	10	14

Note: Total population figures differ from sum of groups because a small percentage of respondents identified themselves as being members of other racial groups, or of combinations of groups. Additionally, the U.S. Census Bureau treats Hispanicism as an ethnicity regardless of race.

high-ground/black-low-ground contention. To answer the second question, of whether black New Orleanians were overrepresented as flood victims, we spatially intersect the same Census 2000 data and the footprint of the Katrina flood. The following analysis was conducted at the metropolitan level (including Orleans, St. Bernard, and Jefferson parishes), and for New Orleans alone (Orleans Parish).

Throughout the metropolitan area, 40 percent of the total population of 988,182 resided in areas that were underwater on September 8, 2005.[34] Blacks outnumbered whites within that flooded area by over a two-to-one ratio, 257,375 to 121,262, even though whites outnumbered blacks metropolis-wide, 500,672 to 429,902. People of Asian and Hispanic ancestry numbered 9,240 and 11,830 among the flooded population, and 25,552 and 49,342 among the total population, respectively. Thus, one in every four whites' homes, one in four Hispanics' homes, and one in three Asians' homes flooded throughout the tri-parish metropolis (24, 24, and 36 percent, respectively), but close to two of every three African Americans' homes (60 percent) were inundated. In sum, whites made up 51 percent of the pre-Katrina metropolitan population and 31 percent of its flood victims; blacks made up 44 and 65 percent; Asians made up 2.6 and 2.3 percent; and Hispanics made up 5 and 3 percent.

Considering New Orleans proper, 61 percent of the total population of 480,256 resided in areas that were flooded on September 8. Blacks outnumbered whites within that flooded area, by over a 3.8-to-1.0 ratio (220,970 to 57,469). But blacks also outnumbered whites citywide *before* the storm, 2.4-to-1.0 (323,868 to 134,012). People of Asian and Hispanic ancestry numbered 7,753 and 7,826 among the flooded population, and 10,751 and 14,663 citywide, respectively. Thus, 43 percent of whites, 53 percent of Hispanics, 68 percent of African Americans, and 72 percent of Asians saw their homes flood in New Orleans. Tallied with respect to pre-Katrina proportions, whites made up 28 percent of New Orleans's pre-Katrina population and 20 percent of its flood victims; African Americans made up 67 percent and 76 percent, respectively; Hispanics made up 3 and 3 percent; and Asians made up 2 and 3 percent.

We can surmise from these statistics that, while African Americans were more likely to be flooded than other groups, the overall racial and ethnic breakdowns of flood victimhood, at either the metropolitan or urban scale, are fairly mixed and not overwhelmingly disproportionate. While any one figure or percentage may be cherry-picked to illustrate a certain narrative, all the race-by-flooding numbers taken as a whole are nuanced and resistant to simple and unconditional generalization. The same may be said of the race-by-elevation relationships presented earlier. Similar statistics come from those killed during the catastrophe: while black victims outnumbered whites by more than double in absolute counts, they comprised 66 percent of the storm deaths in New Orleans and whites made up 31 percent, fairly proportionate to each group's pre-storm relative populations.[35]

Those reports which erroneously implied strong positive correlations between elevation, race, and class—in other words, that higher elevations hosted wealthier residents who were more likely to be white—failed to understand how the perceived technological neutralization of topography originally affected a *negative* relationship between the two: middle-class whites in the 1910s–40s moved enthusiastically into the *lowest*-lying areas, and excluded African Americans with racist deed covenants, even as those areas subsided below sea level. (Note that white, prosperous Lakeview lies significantly lower than the poor, black Lower Ninth Ward.) Additionally, oversimplified reports betrayed a misunderstanding of the role of historical economic and environmental geographies, which explain the otherwise counterintuitive settlement of working-class African Americans along some of the *highest* land in New Orleans—the riverfront. They also failed to recognize that the "pull factors" of suburbia and the "push factors" of the inner city have inspired, in New Orleans and across the nation, a similar out-migration of middle-class non-whites (into areas such as eastern New Orleans) as they had earlier on middle-class whites. In seeking better lives in the suburbs, New Orleanians of all races, classes, and ethnicities, falsely secure in flood-protection and drainage technologies, moved into harm's way in fairly proportionate numbers.

Longitudinal Patterns as a Source of Risk

Migration out of the inner city ended up positioning African Americans, more often than not, on the eastern half of the metropolis, while whites generally settled in the western half. The eastern half of greater New Orleans is generally at more risk of hurricane-induced surge flooding because (1) it is scored and scoured by three major manmade navigation canals (the Industrial Canal, Intracoastal Waterway, and Mississippi River-Gulf Outlet Canal), all of which communicate with the Gulf of Mexico; because (2) it abuts the highly degraded and eroded wetlands where storm-driven Gulf of Mexico surges incur minimum friction as they move inland; and because (3) it lies lower in elevation, being closer to the mouth of the river in this downsloping, prograding delta. The western half of the metropolis, while hardly risk-free, has no major gulf-access navigation canals and enjoys a greater terrestrial buffer, farther geographical distance, and slightly higher elevation from the most dangerous surge paths. (Subsided soils and outfall canals for municipal drainage, which also increase the risk of surge flooding, are found throughout the metropolis.) A glance at a racial distribution map and a Katrina flooding map visually confirms this longitudinal pattern. Calculating the population centroid for whites and blacks throughout the metropolis according to the 2000 Census further corroborates the longitudinal difference: the center-of-balance for the black population lay fully five miles straight east—within a stone's throw of the Industrial Canal, whose floodwalls breached—from the white population's center-of-balance in Carrollton. This longitudinal pattern explains more of the racial variation in the Katrina flooding statistics than elevation-based patterns. In other words, the fact that blacks generally lived on the higher-risk eastern side of the metropolis led to more black flood victims than did their settlement patterns in terms of topographic elevation. Yet, in the press and in popular perception, the latter gets far more attention—much of it weakly substantiated.

Post-Katrina Racial Geographies

Katrina's flood shattered the centuries-old geographies of African American New Orleanians. Most of their circa-2000 population of 324,000 dispersed nationwide by early September 2005, as did other residents. Those who lived in dry areas—particularly homeowners—generally returned by mid-2006 and continued their historical settlement patterns, while those who flooded— particularly renters—continued to face unraveled lives, uncertain futures, and displacement. By summer 2006, fewer than 90,000 black New Orleanians had returned, equaling the city's black population in the year 1910. That figure was contested because of the difficulty of measuring population in a society recovering from a major catastrophe. The 2006 American Community Survey estimated the city's black population at 131,441, still about 60 percent below its pre-Katrina size.[36] Whatever the actual figure, New Orleans's African American population and its total population both increased over the next four years, but at diminishing rates. Total population, which measured around 455,188 before Katrina and plunged to virtually zero in September 2005, rose to an estimated 208,548 on the first anniversary of the storm, to 288,113 by 2007, to 336,644 by 2008, to 354,850 by 2009, and 365,403 by 2010. That last estimation fell short of the actual April 2010 headcount of the federal census, which enumerated 343,829 New Orleanians, or 6 percent fewer than thought. Estimates from 2006–9 can probably be reduced similarly. The percentage of black residents, meanwhile, dropped from 67 percent in 2000 to 60 percent in 2010. An estimated 204,866 African Americans live in New Orleans as per the 2010 census, fully 118,526 fewer than in 2000. The vast majority of the drop happened since 2005 on account of the Katrina flood.[37]

Where do postdiluvian African American New Orleanians reside? In a broad sense, most of the earlier settlement patterns have survived the trauma, although some (such as in Gentilly, Pontchartrain Park, and New Orleans East) persist in visibly thinned dispersions, and one (the severely damaged rear sections of the Lower Ninth Ward) remains largely empty. A major exception involves the public housing projects, which, readers will recall, were undergoing demolition and reconstruction under mixed-income New Urbanism designs as part of HUD's and HANO's HOPE effort. The Katrina flood rendered that controversial redevelopment plan even more polemical amid the rising rents of 2006–7, caused by the large number of flood victims and recovery workers competing for the few habitable housing opportunities. When HUD and HANO proceeded with prediluvian plans to demolish and rebuild the circa-1940s C. J. Peete, St. Bernard, B. W. Cooper, and Lafitte projects, advocates challenged the effort as an attempt to deny poor, displaced African Americans their right to return to the city. Given the housing shortage and large homeless population of the time, their case rested upon the bird-in-hand-is-worth-two-in-bush argument: why destroy sturdy existing housing stock when the promise to redevelop it may not be kept, and when basic financing had not yet been secured? Those favoring the demolition pointed to forty years of deteriorating structural and social conditions as sufficient reason to proceed with HOPE. They also noted that many refurbished HANO apartments had failed to attract tenants, indicating that displaced residents were *not* being denied their wish to return.

Contending that the projects concentrated poverty, incubated social pathologies, and produced intergenerational dependency, the agencies insisted on proceeding with the HOPE concept, although they did agree to stagger the demolition and reconstruction so that some residents could return as work progressed. All that kept the bulldozers from rolling was the approval of the City Council and mayor. The controversy climaxed on December 20, 2007, when the majority-black City Council, amid violent scuffles inside and outside City Hall, unanimously voted to approve the demolitions. Mayor Nagin, himself African American, signed off on the permits, and by 2009, all four complexes lay in rubble, awaiting replacement of their 4,500 units with 3,343 subsidized apartments, 900 market-rate apartments, and another 900 homes for sale.[38] As of 2011, the projects are in various stages of reconstruction—some with financing in question, some with financing secured, others nearly complete and accepting tenants, and all renamed with optimistically chic new monikers.[39] Because the city's public-housing population was and remains about 99 percent black, the eventual success or failure of the HOPE vision—now modified with the Obama administration's new Choice Neighborhoods initiative, which infuses social services, educational opportunities, and employment access to the revisioning—will deeply influence the city's future racial geographies.

Notes

1. Richard Campanella, *Lincoln in New Orleans: The 1828–1831 Flatboat Voyages and Their Place in History* (Lafayette: University of Louisiana Press, 2010). Racial population figures differ somewhat in various aggregations of census data. See also Richard Wade, *Slavery in the Cities: The South, 1820–1860* (London: Oxford University Press, 1964), 326.

2. "Emancipation at the South—Tolerance of Louisiana," *New York Times,* March 28, 1856, citing an article originally published in the *New-Orleans Bulletin.*

3. Ira Berlin, *Generations of Captivity: A History of African-American Slaves* (Cambridge, Mass.: Harvard University Press, 2003), 161–63.

4. James Stuart, *Three Years in North America* (Edinburgh: Robert Cadell and Whittaker and Co., 1833), vol. 2: 241.

5. As quoted by Michael Tadman, *Speculators and Slaves: Masters, Traders, and Slaves in the Old South* (Madison: University of Wisconsin Press, 1989), 95–96. The estimate of 3,000 slaves on the market dates from 1859; a year later, the census enumerated 14,484 slaves residing in the city.

6. Frederic Bancroft, *Slave Trading in the Old South* (Baltimore: J. H. Furst Co., 1931), 312. A survey of newspaper ads revealed that at least 3,500 slave sales occurred in the year 1850 alone, not including unadvertised transactions. Judith Kelleher Schafer, "New Orleans Slavery in 1850 as Seen in Advertisements," *Journal of Southern History* 47, no. 1 (February 1981): 35.

7. "An ordinance in relation to slaves in the city and suburbs of New-Orleans," October 15, 1817, *A General Digest of the Ordinances and Resolutions of the Corporation of New-Orleans* (New Orleans: Jerome Bayon, 1831), 133.

8. Henry J. Leovy, *The Laws and General Ordinances of the City of New Orleans* (New Orleans: E. C. Wharton, 1857), 257.

9. *Louisiana Courier,* May 28, 1831, 4, col. 3 (emphasis added).

10. Interview, Demsey Jordan, by B. E. Davis, 1937, *American Slavery: A Composite Autobiography* (Greenwood Electronic Media), Second Supplemental Series, Texas Narratives, Vol. 06T, 2156 (emphasis added).

11. Larry Ford and Ernst Griffin, "The Ghettoization of Paradise," *Geographical Review* 69 (April 1979): 156–57. See also David T. Herbert and Colin J. Thomas, *Urban Geography: A First Approach* (Chichester, U.K.: David Fulton Publishers, 1982), 312–14.

12. Spatial analysis by Richard Campanella based on data posted by Joseph C. G. Kennedy, *Population of the United States in 1860; Compiled from the Original Returns of the Eighth Census* (Washington, D.C.: Government Printing Office, 1864).

13. Arnold R. Hirsch and Joseph Logsdon, "Introduction: Franco-Africans and African-Americans," in *Creole New Orleans: Race and Americanization,* ed. Hirsch and Logsdon (Baton Rouge: Louisiana State University Press, 1992), 189; David C. Rankin, "The Forgotten People: Free People of Color in New Orleans, 1850–1870," PhD diss., Johns Hopkins University, 80.

14. "An Ordinance to prevent nuisances, and to provide for the security of the public health of the city of New-Orleans," March 18, 1817, *A General Digest of the Ordinances and Resolutions of the Corporation of New-Orleans,* 345.

15. John Adems Paxton, *The New-Orleans Directory and Register,* 1823 (New Orleans: Printed for the Author, 1823), 137.

16. Statistical analysis by Richard Campanella based on population schedules of the 1820 Federal Census, sorted by street and aggregated.

17. Ibid.

18. *Daily Picayune,* "A Kaleidoscopic View of New Orleans," September 23, 1843, 2, col. 3. "Griffe" or "quarteroon" implied a black person with one white grandparent; that is, the offspring of a mulatto and a negro.

19. Based on U.S. Census data graphed by Richard Campanella in *Bienville's Dilemma: A Historical Geography of New Orleans* (Lafayette: University of Louisiana Press, 2008), graphical section.

20. John W. Blassingame, *Black New Orleans, 1860–1880* (Chicago: University of Chicago Press, 1973), 60–61.

21. Peirce F. Lewis, *New Orleans: The Making of an Urban Landscape* (Cambridge, Mass.: Center for American Places, 1976), 46. Black males worked 57 to 60 percent of servant positions in New Orleans in 1870 and 1880, though they comprised 25 and 23 percent, respectively, of the labor force in those years (Blassingame, *Black New Orleans, 1860–1880,* 61).

22. Friedrich Ratzel, *Sketches of Urban and Cultural Life in North America,* trans. Stewart A. Stehlin (1873; translation, New Brunswick, N.J.: Rutgers University Press, 1988), 214 (emphasis added).

23. Hirsch and Logsdon, "Introduction," 189.

24. John Magill, "A Conspiracy of Complicity," *Louisiana Cultural Vistas* 17, no. 3 (Fall 2006): 43.

25. H. W. Gilmore, *Some Basic Census Tract Maps of New Orleans* (New Orleans: Tulane University, 1937), map book stored at Tulane University Special Collections, C5-D10-F6.

26. "Segregation by Co-operation of Civ[i]c Bodies," *Times-Picayune,* November 23, 1924. See also Joel A. Devalcourt, "Streets of Justice? Civil Rights Commemorative Boulevards and the Struggle for Revitalization in African American Communities—A Case Study of Central City, New Orleans," master's thesis, Urban and Regional Planning, University of New Orleans, 2011.

27. Orleans Parish School Board–Office of Planning and Construction, *A Planning and Building Program for New Orleans' Schools* (New Orleans: Orleans Parish School Board, 1952).

28. Daphne Spain, "Race Relations and Residential Segregation in New Orleans: Two Centuries of Paradox," *Annals of the American Academy of Political and Social Science* 441 (January 1979): 82.

29. Lucy Bannerman, "The Big Uneasy: Katrina Exposes Race Divide that Splits American life," *The Herald* (Glasgow), September 3, 2005, 11.

30. "Hard Decisions for New Orleans," *New York Times,* January 14, 2006, Editorial Desk, 14.

31. Susan Page and William Risser, "In New Orleans, 4 out of 5 want to stay there; Blacks report being harder hit by storm than whites," *USA Today,* February 28, 2006, 1A.

32. As quoted by Frank Donze, "Rebuild, but at Your Own Risk, Nagin Says; Recommendations from BNOB Come with Warnings and Worries," *Times-Picayune,* March 20, 2006, 1.

33. Eric Mann, "Race and the High Ground in New Orleans," *World Watch: Vision for a Sustainable World* 19, no. 5 (September–October 2006): 40–42, quoted sub-headline on page 1.

34. Analysis by author using Census 2000 demographic data and flood extent of September 8, representing deeply and persistently flooded areas. "Metropolitan area" is defined here as the contiguous urbanized portions of Orleans, Jefferson, and St. Bernard parishes.

35. Computed by author using "Vital Statistics of All Bodies at St. Gabriel Morgue" (February 23, 2006), update to Louisiana Department of Health and Hospitals News Release, "December 13, 2005: Updated Number of Deceased Victims Recovered Following Hurricane Katrina," www.dhh .louisiana.gov/offices/publications/pubs-192/Deceased%20Victims_2-23-2006_information.pdf (accessed September 2006).

36. Louisiana Department of Health and Hospitals, "2006 Louisiana Health and Population Survey Report—Orleans Parish," Louisiana Recovery Authority (Baton Rouge, January 17, 2007), 3, and U.S. Census Bureau, "American Community Survey—New Orleans City, Louisiana, ACS Demographic and Housing Estimates 2006: Demographic-Sex and Age, Race, Hispanic Origin, Housing Units," factfinder.census.gov (accessed 2008).

37. Calculations based on Federal Census 2000 and 2010 data as reported in "What Census 2010 Reveals about Population and Housing in New Orleans and the Metro Area," by Allison Plyer, Greater New Orleans Community Data Center, April 15, 2011, gnocdc.org/Census2010/index.html.

38. Coleman Warner and Gwen Filosa, "unanimous: Council Votes to Raze 4,500 Units; Old Housing Model to Give Way to Mixed-Income Developments," *Times-Picayune,* December 21, 2007, 1; Rebecca Mowbray, "Sitting in Limbo: Lafitte and B. W. Cooper Top the List of Apartment Complexes Caught Up In a Crunch as Officials Push to extend the Go Zone Deadline," *Times-Picayune,* September 19, 2010, E1–4.

39. For example, the St. Thomas projects became the cheerful "River Garden"; St. Bernard became the pretentious "Columbia Parc at the Bayou District"; C. J. Peete became the Orwellian "Harmony Oaks"; and Lafitte became the chic "Faubourg Lafitte."

Explaining the Unexplainable

Hurricane Katrina, FEMA, and the Bush Administration

ROMAIN HURET

On September 2, 2005, New Orleans mayor C. Ray Nagin called in to Garland Robinette's show on WWL-AM. Robinette had maintained his show even in the midst of one of the worst "natural" disasters in the city's history. After telling Robinette that the federal government should get "their asses moving to New Orleans," Nagin was asked, "Do you believe that the president is seeing this, holding a news conference on it but can't do anything until [Louisiana governor] Kathleen Blanco requested him to do it? And do you know whether or not she has made that request?" Exasperated, the mayor responded, "There is nothing happening. And they're feeding the public a line of bull and they're spinning, and people are dying down here." Nagin's tirade continued for minutes before Robinette could manage to get a word in: "Don't tell me 40,000 people are coming here. They're not here. It's too doggone late," he lamented. "Now get off your asses and do something, and let's fix the biggest goddam crisis in the history of this country." For people from New Orleans listening to such a candid display of the mayor's frustration over the relief effort, it became all too clear just how serious was the breakdown in intergovernmental coordination. On television, striking images of New Orleanians stranded on rooftops and the senseless chaos of looting on Canal Street signaled a man-made disaster. What happened during the last week of August 2005 stunned the public and raised a simple question: "How could this happen *in America*?"[1] During the civic scandal that followed the failure of the Bush administration to alleviate the situation in New Orleans, most people directed their anger at individuals more than systems. However, the governance breakdown was more the failure of a system rather than one of lousy bureaucrats in Washington, D.C.

Actually, George W. Bush meant something serious when he said on September 2, 2005, to the Federal Emergency Management Agency (FEMA)'s terrible director, Michael D. Brown: "Brownie, you're doin' a heck of a job." Brown's job was to follow the standard procedures that had been set up after the terrorist attacks on September 11, 2001. If the management of disasters in the United States has long been intertwined with war-related concerns, the Bush administration strongly reinforced the militarization of federal intervention and changed the priority on the ground. This transformation tends to diminish the long tradition of therapeutic intervention carried out by federal agencies since the creation of the American republic. Whereas, in 1965, Hurricane Betsy transformed New Orleans into a devastated city helped by soldiers from the National Guard, Hurricane Katrina made the city a war zone. The disaster was not the result of irrational and incompetent actors, but the outcome of long and deep-seated reforms of federal and bureaucratic procedures.[2]

As summer vacations drew to an end for millions of Americans at the end of August, the disaster brought them brutally back to the reality of life. Hurricane Katrina was the first major hurricane to hit the country in the age of twenty-four-hour cable coverage. In many ways, television helped construct the frame of meaning in which audiences and decision-makers came to understand it. Essentially, a state of chaos and anarchy was defined. The city was presented as disorganized and on the brink of collapse, less from the storm than from its residents. Rumors of antisocial behavior were particularly virulent, and media coverage facilitated that impression. As New Orleans has always had a reputation as a place of hedonism, there was a short step from the "Big Easy" to the "Big Mess." In the months after the hurricane slammed the city, such frantic coverage still had an impact, and explanations that were offered for what went wrong on the ground in New Orleans focused on the irrational behavior of individuals and institutions.[3]

First and foremost, Michael Brown—a lawyer by training—epitomized the failure of the federal government. This anonymous official of the Bush administration before Hurricane Katrina soon became a household name. And his name was synonymous with nepotism and failure; his emails, sent and received, became famous and were printed on T-shirts. On Wednesday, August 31, 2005, Marty Bahamonde, one of the few FEMA employees to stay in New Orleans, sent him a desperate message about the situation that was "past critical" and about people running "out of food and running out of water at the dome." Brown responded to Bahamonde four minutes later, and his answer became the symbol of his incompetency: "Thanks for the update. Anything specific I need to do or tweak?" The "tweak" answer encapsulated the blindness of federal authorities. Two days later, when confronted by CNN's Soledad O'Brian asking if he was aware of the situation at the Convention Center, Brown again gave a confusing response: "I think it was yesterday morning when we first found out about it. We were just as surprised as everybody else. We didn't know that the city had used that as a staging area. . . . Soledad, I learned about it listening to the news report."[4]

For many Americans, however, Michael Brown was only a pawn in the game of the Bush administration, and soon Brown's critics targeted the president himself and his slow response to the disaster. On the first day of the hurricane, the president was enjoying a long-planned vacation in Crawford, Texas. On Saturday, August 27, from his ranch, Bush declared a state of emergency in Louisiana and authorized FEMA to provide aid, but did not seem very concerned. On the following day, just before the storm struck the city, he made comments on the emergency tucked into longer remarks about the Iraqi Constitution. In the speeches he delivered in Arizona and California during the ensuing days, he integrated remarks dealing with the hurricane. But using add-ons to address the unfolding crisis made him seem little concerned. Not until he was flying back to Washington on Wednesday did he get a glimpse of the disastrous situation. He got that view from his plane, which dipped down over New Orleans. In a news conference in Washington, however, he declared that he was "extremely pleased with the response that every element of the federal government, all of our federal partners have made to this terrible tragedy." Eventually, when *Newsweek* reported that communications adviser Dan Bartlett had to make a DVD of news broadcasts to show the president as he traveled to the area, the impression of an out-of-touch president was cemented. His refusal to admit the failure of federal intervention reinforced the idea that he was insensitive to the people of New Orleans.[5]

Disputes between Governor Blanco and President Bush were seen as another episode of political disorder due to individuals' mismanagement of the situation. When Bush arrived in New Orleans on Friday, September 2, he had a heated discussion with Governor Blanco and Mayor Nagin about the failures of the federal response and the question of who was in charge. At one point Nagin even slammed his hand down on the table and told them, "We just need to cut through this and do what it takes to have a more controlled command structure. If that means federalizing it, let's do it." Governor Blanco refused, and her decision slowed down the process of federal intervention. Despite the crisis being clearly out of control, active-duty troops were not deployed for days because of the standoff between the president and the governor, which continued essentially for the sake of political and personal positioning. Between interviews, with cameras still rolling, Blanco's mistakes were broadcast on national television during a conversation with her press secretary. "I really need to call for the military," she whispered. "Yes, you do. Yes you do," the aide responded dutifully. Blanco then admitted, "I should have started it in the first call," referring to her initial letter to the president. Indeed, the Katrina fiasco was seen as the result of friction between competing bureaucracies and politicians. The unclear crisis response process only made matters worse on all accounts.[6]

As a consequence, for millions of Americans, the disaster seemed man-made, and the men's names were easy to come by—Michael Brown, George W. Bush, and Kathleen Blanco. What made things worse was that other federal employees were not more competent. When the hurricane struck, one journalist concluded: "It appears that there was no one to tell President Bush the plain truth: that the state and local governments had been overwhelmed, that the Federal Emergency Management Agency (FEMA) was not up to the job." As a White House aide noted, "The extremely highly centralized control of the government—the engine of Bush's success—failed him this time." As with the levees of New Orleans itself, there was no backup system in place.[7]

In the weeks that followed, government officials began to point fingers at one another about who was to blame for the slow response. Many members of Congress detailed the incredible complexity of the emergency agencies and strongly deplored "the alphabet soup of coordinating elements" established by the federal government to cope with crisis. Furthermore, congressmen asserted that the American people did not care "about acronyms or organizational charts." In its final report, *A Failure of Initiative,* the U.S. House of Representatives found that the Department of Defense did not incorporate lessons from joint exercises between military and civil authorities that the U.S. Northern Command (NORTHCOM) established after 9/11 for the broad purpose of "homeland defense." In a famous conclusion, it was asserted that if "9/11 was a failure of imagination, then Katrina was a failure of initiative."[8]

In this world without initiative, rumors ran high and increased the sensational portrayal of the disaster. Surprisingly, the rumors were promulgated by officials (including the New Orleans Police Department) and journalists (local newspapers and national television shows such as the *Oprah Winfrey Show*). Many reports were based more on hearsay than on actual observation or eyewitness accounts. There were stories of bodies piled in the Superdome and in basement freezers of the Convention Center. Rumors about violent and animalistic behavior in the shelters spread not only in town, but everywhere in the media. Soon, media were accused of framing blacks as outlaws and savages. The constant replaying of scenes of African Americans "looting" while white residents were looking for food reinforced the notion of a state of social anarchy. A day before evacuations began at the Superdome, Fox News television issued an "alert" warning of "robberies, rapes, carjacking, riots and murders." On television, the disaster was framed as an episode of civil unrest by a violent and uncontrollable crowd of African Americans.[9]

Black activists protested the xenophobic racism and disenfranchisement suggested by the use of the term *refugee* to designate American citizens. Actor Colin Farrell put it bluntly: "If it was a bunch of white people on roofs in the Hamptons, I don't have any fucking doubt there would have been every single helicopter, every plane, every single means that the government has to help these people." After meeting with Louisiana officials, Rev. Jesse Jackson deplored that "many black people feel that their race, their property conditions and their voting patterns have been a factor in the response." Between mismanagement at the top and racism on the ground, the disaster epitomized all underlying fears and racial tensions in the country. Indeed, explanations offered during and in the months following Katrina all reinforce the same irrational analysis based upon individuals' misdemeanors and institutional disorganization. Although all of these explanations are to some extent relevant, a catalyst was needed to create such a catastrophe.

In her song "My FEMA People," veteran Seventh Ward resident Mia X compares New Orleans to countries in the Middle East. The parallel with Baghdad seems obvious, and she deplores the militarization of rescue efforts to help American citizens:

> Ride through my city
> Beirut. Iraq. Ride through my city
> I ride and cry all through the city
> Looking for the culture all through the city
> We were left for dead for vultures all through the city
> It's so much bigger than the weather.[10]

Such a comparison was also obvious for members of the U.S. Congress, who tried to sort things out in the months following the disaster. Furthermore, they explained that "many Americans—and perhaps even some state and local officials—falsely viewed FEMA as some sort of national fire and rescue team." Actually, the disconnect between people's expectations and the agency's new role is a key element to understanding the rational chain of events that led to the catastrophe.[11]

FEMA was founded by President Jimmy Carter in 1979 with two goals: the main one was disaster relief, prevention, and mitigation; a secondary one was coping with civil defense. As Christopher Cooper and Robert Block observe, FEMA was in the first place "an amoeba-like collection of disparate agencies" that inherited many civil defense responsibilities from the Pentagon. Under Ronald Reagan the first goal was neglected and starved of resources, while the secondary one flourished. FEMA set up a "Civil Security Division" with a training center for over a thousand civilian police to handle riots and political disturbances (not disaster relief). This restructuring of FEMA had a catastrophic impact. The response to Hurricane Hugo in 1989 prompted Senator Ernest F. Hollings of South Carolina to characterize the agency as "the sorriest bunch of bureaucratic jackasses I've ever known." The next year, when disasters hit California, U.S. Representative Norman Y. Mineta of that state declared that FEMA "could screw up a two-car parade." At the beginning of the 1990s, it was generally considered an ineffective agency.[12]

FEMA improved remarkably under the leadership of James Lee Witt, an experienced disaster manager appointed by President Bill Clinton in 1993. A high-school dropout who had learned to drive a tractor by the age of six, Witt had worked with Clinton in Arkansas as the director of the state's Office of Emergency Services. He was the first candidate for the FEMA director job to have disaster experience. Under Witt, FEMA not only handled emergency relief well, but set up programs to minimize damage from future disasters. Witt's view of "mitigation" policy included the purchase of vulnerable land to prevent the establishment of settlements. If traditionally the agency arrived after a disaster to provide emergency and relief assistance, Witt asked FEMA's officials to focus more on damage prevention and urged staff members to develop a "life-cycle" model of disaster management. By anticipating the catastrophe, FEMA would not only be more efficient, it would also save taxpayers' money.

When George W. Bush entered the White House, he decided to reorganize FEMA and to reconsider its mission. In the eyes of the new administration, FEMA was a spendthrift agency, a symbol of Clinton populism. The Bush administration tried to cut the federal contribution for large-scale natural disaster expenditures from 75 percent to 50 percent, but Congress balked. Worse still, when a Department of Homeland Security (DHS) was forced upon President Bush by Senator Joseph Lieberman and other Democrats, FEMA lost the cabinet status President Clinton had given it and was folded into the new department. Top personnel left the agency. A survey of eighty-four union personnel found 80 percent saying it was a "poorer agency," and many explained that they would leave if they could get the same salary in another agency. The General Accounting Office rated its morale as one of the lowest of any government agency.

Furthermore, FEMA had to redirect its efforts toward counterterrorism. But when the DHS created a National Response Plan to cope with terrorist attacks, FEMA was not assigned to the task. In 2005, as Christopher Cooper and Robert Block contend, "Homeland Security viewed FEMA as if it were a federal firehouse, where employees flipped burgers until the alarm bell rang." In the Bush administration, FEMA was neither a first-rescue provider nor a coordinating agency. It had to wait until the goals of the National Response Plan had been attained before intervening. The slow response of Brown's FEMA was part of the new national strategic plan envisioned by the Bush administration.[13]

Once President Bush told a group of Crawford, Texas, schoolchildren, "If you know what you believe, decision making is pretty easy." As far as the War on Terrorism was concerned, the "decider," as Andrew Rudalevidge calls him, decided. To support the war on terrorism at home after 9/11, NORTHCOM was established. A National Response Plan was created to manage domestic disasters, incidents of national significance, and catastrophic events. In May 2003, the DHS sent a first draft of this plan to governors, emergency managers, and other first-response providers who were surprised by the way the plan ignored the role local responders played and put the federal government in charge of everything. Some fifteen months and several drafts later, a new version was sent out to the states. A 426-page document, it created a strong bureaucratic apparatus including a "principal federal official" who was in charge of the disaster. Hurricanes were classified as "incidents of national significance." Responsibility and authority, however, were unclear. The secretary of homeland security was in charge, but FEMA's director was supposed to run operations with the Homeland Security Operations Center. However, FEMA's officials were still second-hand players, subordinated to military command. The new chain of command was largely unknown by Americans and especially people from New Orleans who waited in vain for FEMA's buses and rescuers.[14]

On the ground, neither rescuers nor people in general had understood this transformation of FEMA's goals and priorities. More and more people still turned to FEMA officials to provide help to first responders. Director Michael Brown, however, seemed equivocal about assuming total responsibility.

When asked about the slow pace with which FEMA provided ice for people in the city, Brown bluntly answered: "I don't think that's a federal government responsibility to provide ice to keep my hamburger meat in my freezer or refrigerator fresh." Later, even after his resignation, Brown asserted during a congressional hearing that one of the problems was that local officials were "dysfunctional," trying to shift the blame away from the federal government. For many rescuers, it was all the more unbelievable since hurricane plans (such as the Pam Project—a preparedness operation financed in 2004 as part of an exercise by top officials of the DHS) expected FEMA's officials to intervene at the very moment when local rescuers would be overwhelmed by the level of difficulties. During hearings, many state and parish officials said they saw Pam as a "contract" for what the various parties were going to do, and the federal government did not do the things it had committed to doing.[15]

For Brown and his team, federal action was conceived primarily as a safety operation. It came as no surprise that General Russel L. Honoré, commander of Joint Task Force Katrina, once described Katrina as "an enemy that pulled a 'classic military maneuver,' speeding toward land with overwhelming force, surprising and paralyzing the city and countryside and knocking out communications, electricity, water and roads in a 'disaster of biblical proportions.'" Many soldiers in the Arkansas National Guard had just returned from Iraq. When interviewed, they discussed difficulties in assembling the necessary assets (vehicles and so forth) to deploy since so much of their equipment was still being used in Iraq. They also had to deal with crowd-control issues, getting the area ready for evacuation, and then facilitating the evacuation when it finally occurred. The humanitarian role of the army remained unclear for many citizens who stayed in the city. In the documentary *When the Levees Broke* (2009), Kimberly Roberts and her friends went back to see their high schools occupied by soldiers. Soldiers responded to the situation as if citizens were a threat to the safety of the zone. The National Response Plan was, first and foremost, a military plan.

Within five days after the hurricane, the number of National Guard forces had tripled the size of the deployment following Hurricane Andrew. An estimated 63,000 National Guard and military troops were deployed to the Katrina impact region, with most military resources devoted to New Orleans. One soldier reportedly stated that "he didn't think he'd be going back to Baghdad so soon." This increasing militarization of intervention, with the goal of making the disaster area a safe zone, led to a marginalization of FEMA's officials. On July 27, 2005, Dave Liebersbach, the head of the National Emergency Management Association, warned in a letter to Congress that Michael Chertoff's disassembly of FEMA inside the DHS could turn into a national disaster. Thus FEMA's slow and pathetic interventions were not seen as relevant as far as the Bush administration was concerned—it was part of a new strategy to cope with crises and disasters.[16]

Under such circumstances, debates between Governor Blanco and President Bush revolved around these new military priorities and the new view of the Posse Comitatus. "Posse What?" asked one of Nagin's aides when the mayor mentioned the reasons for the institutional conflict. His answer: "It means that the military can't perform law enforcement duties." The Posse Comitatus Act was at the center of the debate. Originally passed on June 18, 1878, as an answer to Reconstruction in the South, the law limits the use of federal military forces within the states as a posse comitatus (that is, "force of the county" in Latin). Since the terrorist attacks of 9/11, the Posse Comitatus Act has been reexamined by conservative officials. In February 2002, Colonel John R. Brinkerhoff asked for reinforcement of federal power in case of crisis: "President Bush and Congress should initiate action to enact a new law that would set forth in clear terms a statement of the rules for using military forces for homeland security and for enforcing the laws of the United States. Things have changed a lot since 1878, and the Posse Comitatus Act is not only irrelevant but also downright dangerous to the proper and effective use of military forces for domestic affairs." A former FEMA acting associate director for national preparedness from 1981 to 1983, Brinkerhoff argued that there was a misunderstanding about the actual meaning of the Posse Comitatus Act. It was not enacted to prevent members of the military services from acting as a national police force. Instead, it was enacted to prevent the armed forces from being abused by having soldiers pressed into service as police officers (a posse) by local law enforcement officials.[17]

According to Brinkerhoff, the Posse Comitatus Act had become far removed from its original purpose in its historical context, considering that "the intent of the act was not to preclude the Army from enforcing the law but instead to allow the Army to do this only when directed to do so by the President or Congress." For the Bush administration, after the hurricane slammed the city, this interpretation was an opportunity to allow the intervention of the federal government. If Governor Blanco and President Bush disagreed during the meeting inside Air Force One on September 2, it was not only about political prerogatives, but more fundamentally about the role and goal of officials on the ground. Discussions centered on the very rationality of the plan designed by Bush's administration to cope with terrorist attacks.[18]

Historically, in the United States, responsibility for dealing with disaster response is located at the local level. What makes Hurricane Katrina unusual is that many people expected that federal resources would be made available to be used by local and state officials. During congressional hearings, a citizen of New Orleans, Ann Thompson, summed up this high level of expectation among citizens who stayed in the city: "We were abandoned. City officials did nothing to protect us. We were told to go to the Superdome, the Convention Center, the interstate bridge for safety. We did this more than once. In fact, we tried them all for every day over a week. We saw buses, helicopters and FEMA trucks, but no one stopped to help us. We never felt so cut off in all our lives. When you feel like this you do one of two things, you either give up or go into survival mode. We chose the latter. This is how we made it. We slept next to dead bodies, we slept on streets at least four times next to human feces and urine. There was garbage everywhere in the city. Panic and fear had taken over." Such an emotional statement reveals the extent of the moral, social, and political disaster. It also displays a misunderstanding of the new management of disasters after 9/11. After the reorganization following terrorist attacks, FEMA lost its independent status and had to find new bureaucratic prerogatives among twenty-two agencies located in the new Department of Homeland Security. Its emergency mission had been reconfigured to become first and foremost a security mission according to the directive of the National Response Plan. In other words, the federal response to Katrina was the result of rational choices, made by U.S. officials to cope with issues of the country in the twenty-first century.[19]

Notes

1. For the full transcript of Nagin's comments, see "Mayor to Feds: 'Get Off Your Asses,'" September 2, 2005, www.cnn.com/2005/US/09/02/nagin.transcript/; Robert Tice Lalka, "The Posse Comitatus Act and the Government Response to Hurricane Katrina," *The Current* 11, no. 2 (Spring 2008): 81–99; Sally Forman, *Eye of the Storm: Inside City Hall during Katrina* (Bloomington Ind.: AuthorHouse, 2007); Walter Brasch, *"Unacceptable": The Federal Response to Hurricane Katrina* (Charleston, S.C.: BookSurge, 2005); Michael Eric Dyson, *Come Hell or High Water: Hurricane Katrina and the Color of Disaster* (New York: Basic Civitas, 2005).

2. Kevin Rozario, *The Culture of Calamity: Disaster and the Making of Modern America* (Chicago, University of Chicago Press, 2007); Michele Dauber, "Fate, Responsibility, and 'Natural' Dis-

aster Relief: Narrating the American Welfare State," *Law and Society Review* 33 (1999): 257–318; Michele Dauber, *The Sympathetic State: Disaster Relief and the Origins of the American Welfare State* (Chicago: University of Chicago Press, 2013); Marie Salwa, "L'Ouragan Betsy: Perspectives de Recherché," Mémoire de Master, École des Hautes Études en Sciences Sociales, 2007; U.S. Congress, *The Hurricane Betsy Disaster of September 1965* (Washington, D.C.: U.S. Govt. Printing Office, 1965); *Hurricane Betsy, 1965: A Selective Analysis of Organizational Response in the New Orleans Area* (Columbus: Ohio State University, 1979).

3. *Economist,* September 10–16, 2005; *New Republic,* September 19, 2005; *Time,* September 19, 2005; *U.S. News and World Report,* September 19, 2005.

4. "Hurricane Katrina Document Analysis: The E-Mails of Michael Brown," staff report for Rep. Charles Melancon, U.S. House of Representatives, November 2, 2005, katrinacoverage.com/2005/11/03/charlie-melancon-and-michael-browns-emails.html; e-mail from Marthy Bahamonde to Michael D. Brown, Aug. 31, 2005, katrinacoverage.com/2005/11/03/charlie-melancon-and-michael-browns-emails.html; quotation in "City of New Orleans Falling Deeper into Chaos and Desperation," September 2, 2005, transcripts.cnn.com/TRANSCRIPTS/0509/02/ltm.01.html; House Select Bipartisan Committee to Investigate the Preparation and Response to Hurricane Katrina, *Testimony of Michael Brown, Hurricane Katrina: The Role of the Federal Emergency Management Agency, 109th Congress, 2005* (Washington D.C., U.S. Government Printing Office, 2006), 116.

5. Martha Joynt Kumar, "Managing the News: The Bush Communications Operation" in *The Polarized Presidency of George W. Bush,* ed. George C. Edwards III and Desmond S. King (New York: Oxford University Press, 2001), 379–80; Kumar, "Managing the News," 380; Thomas Evans et al., "How Bush Blew It," *Newsweek,* Sept. 15, 2005.

6. Brasch, *"Unacceptable";* Dyson, *Come Hell or High Water.*

7. Quotations in Andrew Rudalevidge, "'The Decider': Issue Management and the Bush White House," in *The George W. Bush Legacy,* ed. Colin Campbell, Bert A. Rockman, and Andrew Rudalevidge (Washington, D.C.: CQ Press, 2008), 154.

8. *A Failure of Initiative, Final Report of the Select Bipartisan Committee to Investigate the Preparation for and Response to Hurricane Katrina, U.S. House of Representatives* (Washington, D.C.: U.S. Government Printing Office, February 15, 2006), x, xi.

9. Dyson, *Come Hell or High Water.*

10. Zenia Kish, "'My FEMA People': Hip-Hop as Disaster Recovery in the Katrina Diaspora," *American Quarterly* 61, no. 3 (September 2009): 676.

11. *A Failure of Initiative,* 13.

12. Jeffrey K. Stine, "Environmental Policy during the Carter Presidency," in *The Carter Presidency: Policy Choices in the Post–New Deal Era,* ed. Gary M. Fink and Hugh David Graham (Lawrence: University Press of Kansas, 1998), 179–201; Christopher Cooper and Robert Block, *Disaster: Hurricane Katrina and the Failure of Homeland Security* (New York, Time Books, 2006); quotation in Charles Perrow, "Using Organizations: The Case of FEMA," understandingkatrina.ssrc.org/Perrow; National Academy of Public Administration, *Coping with Catastrophe: Building an Emergency Management System to Meet People's Needs in Natural and Manmade Disaster* (Washington D.C.: Diane Publishing Co., 1993).

13. Jon Elliston, "Disaster in the Making," *Independent Weekly,* September 22, 2004, www.indyweek.com/durham/2004-09-22/cover.html.

14. Rudalevidge, "'The Decider,'" 155.

15. Quotation in Eric Lipton and Shane Scott, "Leader of Federal Effort Feels the Heat," *New York Times,* March 9, 2005, 1; Cooper and Block, *Disaster,* 3–22.

16. Quotation in Kathleen Tierney and Christine Bevc, "Disaster as War: Militarism and the Social Construction of Disaster in New Orleans," in *The Sociology of Katrina: Perspectives on a Modern Catastrophe,* ed. David Brunsma, David Overfelt, and J. Steven Picou (Lanham, Md.: Rowman & Littlefield, 2007), 42.

17. Forman, *Eye of the Storm,* 154; Lalka, "The Posse Comitatus Act and the Government Response to Hurricane Katrina," 81–99.

18. John R. Brinkerhoff, "The Posse Comitatus Act and Homeland Security," February 2002, www.homelandsecurity.org/journal/Articles/brinkerhoffpossecomitatus.htm.

19. Romain Huret, *Katrina, 2005: L'Ouragan, l'Etat et les Pauvres* (Paris: Editions de l'EHESS, 2010).

Picturing the Catastrophe

News Photographs in the First Weeks after Katrina

JEAN KEMPF

Catastrophe: 1. The final event of the dramatic action, esp. of a tragedy. 2. a momentous tragic event ranging from extreme misfortune to utter overthrow or ruin. 3. a violent and sudden change in a feature of the earth. 4. utter failure.
  —*Webster's Dictionary*

Like all contemporary events, Hurricane Katrina was a news event in which images—and particularly still images—played a defining role. The narratives constructed around the event work as a system of explanations, be they mythological, biblical, or sociological. First among them, the paradigm of the naming of the occurrence itself (catastrophe, tragedy, and so forth) immediately proposes a set of definitions that attempt to domesticate and make sense of that which makes no sense, by placing the idiosyncratic nature of the event within a known paradigm. Photographs are one of the major elements of this characterization.

Hurricane Katrina was visually reported in still photographs published in various news media.[1] The way Katrina was constructed through images for national and international audiences as well as for future memory bears a striking resemblance to other natural or man-made catastrophes. The codes of representation at work in this event were elaborated almost eighty years ago. For all intents and purposes, images of Katrina in the print media—the case of the internet is much fuzzier—were in no way different from how the *Ur* catastrophe of American culture in the twentieth century—the Great Depression—was portrayed in picture magazines and syndicated newspapers of the time. This continuity evidences the permanence of a very few deep rhetorical visual figures since at least the modernist period. Despite major changes in communication with the advent of television and digital media, our visual world and consequently our perception of events have remained remarkably stable. One can even witness the return of the still image in the public sphere since the early 2000s after decades of the prominence of television, a reversal due to the development of amateur digital photography coupled with the internet.

Of course, the proposition could be reversed. The continuity may be primarily political—and not visual—that is, ever since the 1930s, information providers and "shapers" (journalists and editors) in all liberal democracies have merely reinforced the dominant ideology of social engineering. Whatever the causal relationship, however, there is a true historical continuity in the belief in showing (rather than telling) and particularly in showing through photographs.[2]

The Ambiguous Image of Katrina

Despite the difference in the particulars, the visual rendering of Katrina, both of the hurricane and the flood, bears a striking resemblance to previous treatments of catastrophes. Nature—its magnificence but also its vagaries—is central to American life for reasons both historical and geographical. The awe that nature inspires—from giant trees or geysers to tornadoes and hurricanes—has blended with the romantic sensibility as well as the more pragmatic, positive, can-do mentality that became a landmark of the famous "pioneer spirit" and part and parcel of the American experience. The sheer violence of natural phenomena has structured the American psyche. Many Americans live in danger zones, risking instant annihilation, from the Florida Keys to California, and they know it. This land is their land, however, and they even draw a particular if paradoxical pride—almost a form of identity—from this situation which perpetually challenges their power to remold the land and shape it for their exclusive use. What is at stake is not so much "making do with nature" as "bending it to man's use" as the ultimate gauge of success, and thus of true humanity. The history of the conquest of American soil is fraught with such instances. "Natural disaster" images cannot be read without bearing in mind that such disasters are part of American cultural history as natural *challenges*.

Katrina must thus be seen within a long series of catastrophes. One can mention the San Francisco earthquake, photographed by Arnold Genthe, whose images were largely publicized; the Galveston hurricane of 1900; the great Mississippi floods of 1927; the Tri-State tornado of 1925; the great New England hurricane of 1938, which hit a part of the country not used to such damage; and the great blizzard of 1993. Katrina, however, was not simply a giant hurricane but (mostly) a catastrophic flood, and the specific place of floods in human mythologies—many cosmologies begin with one—gave it a special place in imagination. Yet, perhaps the model event was paradoxically the Great Depression because images played a central part in depicting the *economic* crisis as a *natural* catastrophe and also deeply shaped the iconography of victims. Much of the New Deal message of recovery was also predicated on the innocence of the affected populations.[3]

For American audiences of the late twentieth and twenty-first centuries, however, the largest source of traumatic visual references is to be found in disasters such as floods, landslides, and explosions taking place overseas and, most significantly, affecting nonwhite populations. This made the "normal" victims of a catastrophe non-Americans and nonwhite, a fact that deeply shaped the perception of Katrina. New Orleans thus became actually caught up in its very image: with its exotic flavor in American culture, the city that is a little of the Caribbean culture on American soil—and is marketed as such—was made a real third-world city by Katrina and thus severed from the mainland, as it were.

Making Pictures Is a Commitment

All the first images and many of the most striking ones were made by local news photographers who were themselves hit by the catastrophe in their own lives. They kept working despite the immediate danger to themselves and their families: "You're not covering a story, you're covering your life," says one.[4] Conversely, the total disruption of communication turned the reporters in the field into free agents relying on their own feelings, knowledge, and capacity to make decisions and to band together for greater safety.[5] For them, this moment marked a truly liberating experience from the highly formatted practice of journalism in the days of instant communication, and allowed them to experience again some of the imagined (or real) freedom of the great mythical figures of the 1930s–50s, the documentarians and war reporters who shaped the field. In other words, they regained "agency" even though sometimes only in fantasy.

Photographers describe their total involvement in the cause of reporting as a commitment to the community, giving real meaning to their ordinary jobs in words that make them soldiers of information but soldiers who would paradoxically only obey their own conscience. Clearly, in these moments, the individual photographers felt the embodiment of a whole profession. In so doing, they transformed their contingent individual selves into emblems, a process parallel to what they do in their own photographs.

More prosaically, this exceptional event was also a chance to break into the national or even international news market, to make a name for oneself, to possibly "make a difference."[6] The Katrina photographers, most of them with only local experience, probably weaned on tales of war photographers, got a chance to experience something close to working in a combat zone, and akin to real adventure—a term used several times in the accounts. As veteran photojournalist Bill Pierce expressed it, "They are going to get an education that will change them forever."[7]

The most important lesson they encountered involved the fundamental journalistic question of choosing between being a witness or a player. The *New Orleans Times-Picayune* photographers, as well as others, repeatedly said that, given the extent of the flood and the limited number of rescuers, they had to choose between rescuing people and taking their pictures. They tried, it seems, to mitigate the two, and often the documents tell us that the issue was positive for the people concerned.[8]

Times-Picayune photographer Ted Jackson raises similar moral issues when he describes the complete anarchy that took hold of New Orleans in the first two days: anything went, whatever you needed you just took, a situation that he felt was both exhilarating (the complete lifting of all of society's pro-hibitions) and scary (the end of one's status as a protected individual, a white man and a journalist). For him, it was a moral proving ground that served as a metaphor for the larger ethics of reporting, the honesty that is the foundation of any reliable information for a democratic society.

Even more significant than the stories themselves is the tone of the tales told by fellow journalists meeting with the photographers who operated in New Orleans during the worst part of the crisis. The lengthy personal cap-tions appended to many images and narrating the experience surrounding the taking of the image—not just presenting data on the subject as is usually the norm—reveal adventure, everyday humanism, and honesty (truthfulness) as befits true heroes of the communication age.

Framing the Scene, Creating a Spectacle

The power of the photographic image rests largely on the photographer's word: if the belief in the truth-producing power of the mechanical image is now long gone, the validation attached to the "sources" is still one of the main truth-tests for photos. Showing a photograph remains an act of truth-telling predicated on the assumptions that someone (the photographer) was there, that this someone did not fundamentally alter or distort the "moment" by his or her selection, and that no such process took place in the editing chain.[9] The literature on doctored images, or simply on "misleading images," is large, and digital technology has only increased the possibilities for manipulation. These developments have placed even greater responsibility on the publication chain, thereby reinforcing the necessity of the trustworthiness of its very first link, the photographer.[10] The photographers' reactions to the scene are conditioned by three factors: cultural background/training, technology, and access.

Access is one of the often disregarded and yet paramount conditions of visibility. In the context of war, it is often discussed in legal and political terms. Yet physical access and the impediments to it are nearly as important. In the case of Katrina, certain parts of the city were easier to reach than others, and certain places concentrated a large number of people (the Superdome in particular), making them choice sites for images. Visual evidence is thus more the result of availability of access than of real statistical evidence, although the two are not necessarily opposite. Similarly, aerial views—a classic of flood images—although dependent on the availability of air transport for journalists, are used not only because of their telling function, but also as a direct result of access problems in disaster zones.

Then comes rhetoric. In order to be "catchy"—the mantra of all photojournalism—a picture must not only show things (telling the story differentiates art and journalism) but also encode them within a context of visual references. In other words, pictures need to be constructed according to certain visual principles designed to capture the viewer's attention: contrast, opposition, composition. Their simplicity and repetition over time produce a remarkable stability in the treatment of shock scenes in press photos.

Technology has allowed considerable progress in the implementation of old formulas, making possible what was inaccessible to previous generations: zoom lenses going from very wide angles to very long telephotos, highly sensitive sensors allowing most pictures to be taken without flash even in low-level lighting. Scenes thus appear to be more natural and less intrusive than those taken by earlier documentarians such as Riis, Hine, FSA, or Weegee, and the chiaroscuro effect easily associates these images with art. Not surprisingly, the corpus of images made by aspiring photographers from the University of Texas School of Journalism is more heavily marked by "artsy" compositions and lighting, as they often prefer the pictorial reference to the explicitly narrative form, a feature reinforced by the presentation of some of their images in black-and-white.[11] Many of the frames in this "junior" series are askance, evidencing a constructivist aesthetic emphasizing the formal visual composition. In the massive output of professional images on Katrina, "medium range" shots are rare; 80 percent of the photos are either wide-angle shots or long-range shots which increase the narrative power of the image.[12]

The long-range shots allow abstraction and focus by selection of a small part of a scene and by a short depth of field. They work the way close-ups do. The old woman's hands in the Midwest by Russell Lee or the sharecropper's boots by Walker Evans have become a case in point in the history of photographic rhetoric.[13] In photographs, the detail is more than itself: it becomes a metonymy. In the case of Katrina, the shot of a man's feet wearing makeshift shoes reading "Keep moving" combines many of those qualities of a powerful documentary icon (see fig 3.1).[14] On the other hand, the wide-angle lens places a subject in the foreground against the background of the scene, thus

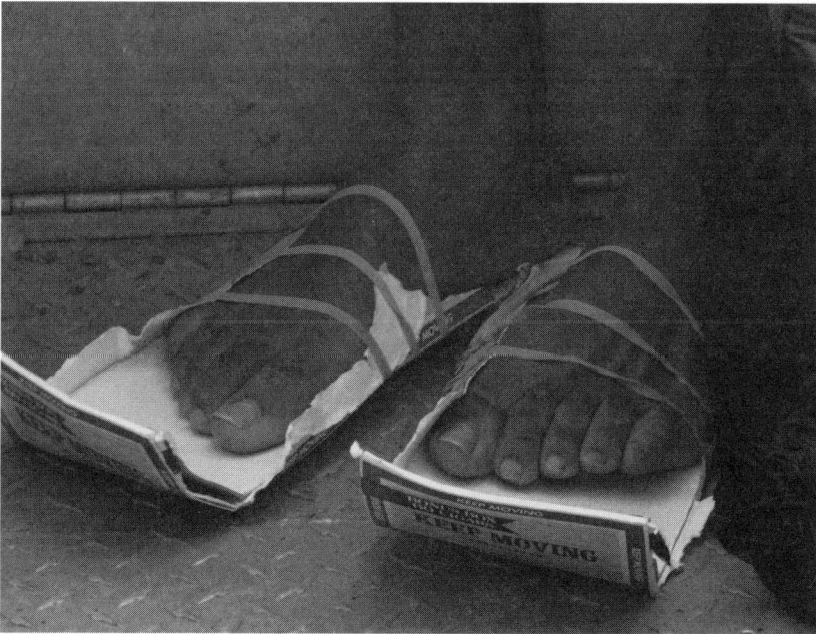

Fig. 3.1. Jeremiah Ward wears makeshift shoes after being rescued in the Ninth Ward on August 30, 2005. Photo by Irwin Thompson, *Dallas Morning News*.

producing at the same time contextualization and dramatization. It is used to describe the chaos resulting from the hurricane/flood, with the surrealistic superposition of everyday objects (in a broken or tumbled state) relocated in unusual/absurd or especially pathetic locations.[15] This classic visual rhetoric was established in the 1930s by such diverse photographic movements as the Russian constructivists, the German *Neue Sachlichkeit,* and the American Farm Security Administration.[16] Despite their sad message, these images make the viewer experience the "evil beauty" of such chance meetings or superimposition, along with the terrible power of nature—in other words, experience the feeling of the sublime.

The last rhetorical figure is the aerial shots.[17] Aerial shots are classics of the floods, even of the early ones, such as the 1927 Mississippi flood. Although not numerous in the Katrina files I consulted, they were published by all media.[18] Visually they show how formerly well-known places were made unrecognizable and New Orleans transformed into a sinking ship or a Venice of the Gulf.[19] Eventually aerial shots turned the city into painterly abstractions. This is particularly true of one shot published in *Time* that cuts out the horizon, dislocates the scene geographically, and makes it into a pattern of abstract ob-

Fig. 3.2. Water surrounds homes just east of downtown in this aerial view of damage from Katrina on August 30, 2005. Photo by Smiley N. Pool, *Dallas Morning News.*

jects slowly realized by the viewer as being roofs (see fig. 3.2). Beyond its journalistic meaning (only accessible to locals or through detailed explanation of
the evacuation procedures), a flooded parking lot full of schoolbuses provides
both the pleasurable rhythm of repetition of the famous yellow boxes and the
sadness of the drowning of an icon.[20] If the destructive force of the hurricane
could be compared in visual effects to that of the 9/11 attack, which produced
a holocaust, the flood itself—which was the real Katrina disaster—was an essentially static event. And yet, from such a static event, photographers constructed a spectacle.

The Rhetoric of Disaster

Images of Katrina—and in fact of most disasters—can be roughly divided
into three categories: statements (what the hurricane and the flood did), action (rescuing and taking over from nature), and knowledge (explaining the
causes and the effects and preparing the aftermath). Though the volume of
published images in each category changed through time and differed according to publications, the rhetorical figures remained stable.

Effects

Catastrophes—whether lethal or not—are most often described by the traces
they leave, to make the viewers understand the nature of the event through
metonymic reasoning displacing the focus from the event itself to the victims.
A catastrophe is first and foremost defined by its victims. Chronologically this
was the first and the longer-lasting theme, and can be characterized by three
types of images: the effect on objects, the effect on bodies, and the effect on
homes.

The tracing of objects is first and foremost the displacement and the surrealistic conjunction of objects, when the flotsam and jetsam come to be seen.
This is exemplified in the feature section of the *Times-Picayune* (September
15, 2005, 12), called "Left Behind," which chronicles a series of abandoned objects as pathetic traces of the flight, but also of lives lost.[21] The feature has a
three-little-pigs effect with houses literally blown away, warped iron structures, trucks ending up in trees and boats on highways (*Times-Picayune,* September 11, 2005, 4). *National Geographic* in its August 2006 edition even offered an anniversary panoramic center-spread called "before and after" on
the Gulf Coast (Holly Beach). It focused on the devastating power of the hur

ricane as well as on the Lower Ninth Ward which, according to a caption, "remained a scene of apocalypse and a reflection of lives thrown into chaos"
even six months after "Katrina splintered the modest houses and wrecked the
cars." The *Times-Picayune* (September 1, 2005, 15) sums up the situation: "Unbelievable debris, unbearable sadness, unrelenting need for water are among
the things Hurricane Katrina deposited in Gulfport, Miss." The caption points
explicitly to the rhetorical figure used in the images: the debris represents
what cannot be represented—the deep, long-term consequences of a disaster. How does one show broken lives except metaphorically by showing the
broken objects of those lives? Then come the antidotes: the reconstruction of
communities: a broken house should not mean a broken home.

The most powerful *locus* is undoubtedly the home. It draws its evocative
power from its mythical place in society but also from playing the role of both
metaphor (broken house means broken life) and metonymy (the house as
container of people). Often the roof (the synecdoche of house and home) was
the only road to salvation. But most of the time it is the intimate, the personal,
the cherished, and of course the "duly and difficultly earned" that are washed
away by the catastrophe. The scene is tainted by the water, and images of
return (appearing as early as September 6, 2005, in the *Times-Picayune*) are
sometimes treated in a hopeful way ("life's back and we're going to fight") but
most often in an elegiac mood as in the report *National Geographic* made for
Katrina's first anniversary.[22] Almost in anticipation, in its August 2005 issue,
National Geographic had published a striking (and rhetorically perfect) picture of the effects of hurricanes in Florida on a middle-class girl of nineteen,
whose life had been "erased" by Hurricane Ivan.[23] In August 2006, *National
Geographic* featured two images of disaster, combining most of the elements
used in the rhetoric of loss. One is that of a child's suit and shirt stained with
mud and mold and the other of a closet in much the same state of disrepair.[24]
These are the closest metaphorical representations of death one can come to,
all the more poignant as they use the image of the child, a figure of helplessness and innocence.

The bulk of the representation of suffering, however, is borne by faces,
preferably of women, children, and elderly people. This old rhetorical choice
goes back to the early days of the use of photography in social reform and was
greatly developed by Farm Security Administration photographers. It draws
on classic Christian iconography, as shown by Alan Sekula.[25] Most of the time
Katrina victims are seen from above, shot with a wide-angle lens, increasing
the effect of being crushed by events, or even with their faces hidden in their

hands. One exception is the image of the evacuation of one handicapped girl and an elderly man from the Superdome published in the *Times-Picayune* (September 1, 2005, 14), deliberately taken at a low angle, probably more as a way of contextualizing the evacuation (to show the dome in the background) than as a way of magnifying the subjects. Although they are identified by name to conform with modern journalistic convention, as opposed to early twentieth-century fashion which identified people by class or type, the effect on viewers remain the same. The face is supposed to tell the story, and the individual expression becomes the expression of a whole class of people. As Lange's California pea picker was the Migrant (Mother) and embodied the plight of migrants, certain faces became the faces of Katrina, especially because they were used by magazines on their covers and inside full-page spreads.[26]

Aimed at audiences outside the area, and for the most part international in scope, news weeklies played their tune of borderless double-page spreads, punctuated with the kinds of words and phrases used in such circumstances, like "tragedy," "how did it happen?" and "rebuilding a dream."[27] As there are relatively few images in each issue, these are turned into emblems, using simple symbols as well as the reservoir of figures constructed over time by disaster reports. Among those is the classic image of the Madonna (Virgin Mary with child, secularized as mother with child) as seen in *Time* magazine with its opening image (September 12, 2005, issue), or *Paris Match* using the same picture for its cover or inside pages (September 8–14, 2005, issue), or a similar one in *Newsweek* (cover of September 12, 2005, issue). Variations on the theme were used by *Newsweek* for the cover of its second Katrina edition (September 19, 2005) showing an extreme closeup of a black child crying (cover of September 19, 2005, issue);[28] and in an almost unique front page, the *Times-Picayune* of September 2, 2005, with Angela Perkins screaming "Help us, please!"[29] The elderly were used as well, particularly in one instance, by *Paris Match* and *Newsweek* in a criticism of the reaction of the federal government towards the population, by showing an elderly woman draped in an American flag with the captions: "America at half-mast" ("L'Amérique en berne," *Paris Match,* September 8–14, 2005) and "The Other America," *Newsweek* (September 19, 2005).

Action

The rescue actions were chronicled alongside another type of action—which came to be as much publicized as the rescue—the policing and even militarizing of New Orleans. The connection between those two aspects of the "intervention" can be seen with the army helicopters and trucks, half-military, half-humanitarian. A largely black New Orleans was shown as an overseas theater of operation, one more zone where American forces projected themselves to implement the Pax Americana, but hardly as a part of the national territory.

Police officers and other officials were first presented as saviors of lives, as on the front page of the *Times-Picayune* on August 30, 2005, where two heavily equipped policemen save an elderly man from drowning in his own house.[30] Helmets, boots, radios, high-visibility vests are the emblems of this job, and boats are lifelines. Then came the medical aspects of rescue—with a few scenes seemingly drawn from *ER* slowly turning into a third-world infirmary as help did not arrive. On the fifth day help came at last, and it did so from the sky with big choppers—the staple of rescue technology since the Korean War—as well as heavy army technology. But then the action completely changed meaning, or to put it visually, the scene moved from *M*A*S*H* and *ER* to *24 Hours*. Special forces in full gear, coming out from the well-established iconography of action movies and series and from the war in Afghanistan or Iraq, descended on the city.[31] New Orleans thus became the first part of the American territory to be invaded by Homeland Security forces, treating the civilian population as (potential) enemy aliens and Louisiana as non-U.S. territory. The political effect of action visuals was ambiguous at best, at worst clearly negative for the federal government. The watershed moment seemed to take place on September 8, 2005, when mandatory evacuation was decided and made the front page of the *Times-Picayune* in extremely clear words: "Clear out or else," with a picture of a Black Hawk helicopter hovering above a black man paddling home in the opposite direction, transforming a search-and-rescue mission into a search-and-destroy mission.[32]

Lastly, in disaster communication, among the major figures of action alongside active rescuers are the iconic figures of politicians, high officials, or even entertainment stars visiting disaster areas. They appear on the scene, or its whereabouts, to signal that society and its solidarity functions are present. The attacks on 9/11 were a catastrophe of a different nature, which obviously called for a prominent visual place for the commander-in-chief.[33] But ever since the trope of war was established by Franklin D. Roosevelt in his first inaugural address, the president of the United States and to some extent governors have clearly established the link between economic/social/political problems and natural disasters. All come under the chief executive's remit, and emergency action to protect or relieve American citizens is part and parcel of his or her duty. But here, in both local and national or international coverage, George W. Bush was prominently invisible. The first image of him is "out-of-scene" and hardly shows his face; the second one is of Marine One (the president's helicopter) flying low over New Orleans; the third one is of a meeting with firefighters and a dog that was saved from the rubble on September 12; and when finally G. W. Bush makes it to the front page, the portrayal is rather terrible for him, as well as for politicians in general, as one cannot avoid reading the image metaphorically: all duck to avoid a low hanging wire; Governor Blanco and Mayor Nagin look terrified while President Bush almost disappears in the command car (only General Honoré seems unimpressed).[34] In any case, this was quite a different picture from the 9/11 imagery when a fighting Bush was standing on top of the rubble, with the rescuers, encouraging them like a nineteenth-century general in the midst of battle.[35] As for Governor Blanco, she is seen on the verge of a nervous breakdown, so tired that she leans on the president for support, a rather ironic image as their relationship was less than good.[36]

Despite the actual coverage in articles, Katrina appears *visually* as the demise of politicians and as the collapse of civil and political society and ultimately of democracy in its protective function. Faced with elemental forces, people seem abandoned in the city after the first evacuation, able to count only on their own strength and on others as affected as themselves. Pictures describe the emergence of this person-to-person solidarity, of community building amongst the ruins, reenacting the origins of the peopling of the nation. The description of the killer wind, followed by the inexorable flood and eventually the vengeful fires (*Times-Picayune,* September 7, 2005), as well as "diseases" point at a mythical reading of the catastrophe, a plague, mobilizing the four elements. One *Time* double-page even connects it to the burning of Atlanta, a deeply engrained mythical holocaust in southern consciousness.[37] One does not have to subscribe to extreme interpretations of the causes for the disaster[38] to see the catastrophe portrayed as an American myth: the *Times-Picayune* image announcing the fires and diseases on September 7, 2005, featured a Christ-like figure (wearing a visible gold cross on his chest) walking out of a primeval fog.[39] In other words, by obliterating civilization and creating a complete disruption in the traditional taken-for-granted systems, Katrina threw New Orleans back into a state of lawlessness characteristic of frontier days, those of a "generic" frontier, of certain parts of the American West and the Rockies, with guns in the streets and mud everywhere.[40] In such historical/mythical reading, bridging all geographical and chronological gaps, the refugees in the Superdome, or in other relief centers, could be compared to those huddled masses, the refuse on the teeming shores of massive nineteenth-century immigration—images revived in recent years by the pictures of Mexican immigrants trying to cross the border. In fact what was at stake in this visual confusion was a strangely contradictory message: on the one hand a form of ontogenesis of America (as if each generation had to make the experience of the founding of the nation again), on the other hand the visual demonstration that New Orleans's citizens—and more generally those of Louisiana—were aliens.

Explanations: Science and Politics

Direct human responsibilities in this "natural" catastrophe were indicated by three types of publications. First were those with a more liberal leaning—*Newsweek* for instance, which devoted its second Katrina issue to the race factor while *Time* simply saw a "system failure."[41] Second were those with an environmentalist agenda—*National Geographic* and to some extent *Time*, which focused its third Katrina issue on the human impact on hurricane strength. Finally, there were foreign newspapers and magazines, which emphasized the failure of George W. Bush in dealing with domestic problems and used the well-known metaphor of the "giant with clay feet," pointing to America's failure to protect its own citizens and to the unfair nature of American society.[42]

The environmentalist approach brings no surprise and concentrates mostly on views from space of the hurricane and its maelstrom effect, with its massive otherworldly white texture reminding the viewer of supernovas or the surface of Jupiter. *Time* chose a different metaphor—and a clever layout—transforming the image of a wave into that of a hurricane (with its rotating pattern), invading the reader's space by jumping out of the frame as it were and eating part of the title of the magazine.[43]

One of the charges made was the unequal toll of this natural disaster on residents according to their color and social class. The second issue of *Newsweek* on the topic (September 19, 2005) editorialized on the race factor by placing it on its cover: "Poverty, Race and Katrina: The Lessons of a National Shame." The choice of cover images by *Newsweek* made explicit references to wars and disasters in developing countries (September 12 and 19, 2005, issues). The visual effect, however, was two-sided. While the large visible presence of African Americans was a mere translation of a demographic fact, their visibility made this *American* disaster a *black* disaster, racializing the issue in an almost mechanical way. *Paris Match,* while acknowledging the racial fact, ends with a personal portrait of a white woman and her two children in Biloxi with a caption that reads, rather vaguely: "Black or white, those who suffered most were the poor," thus drawing a rather French conclusion on an American situation: the real fracture line for *Paris Match* is social, not racial.

Paris Match gives a good example of the indictment the United States received for the way the country treats its population. This does not occur, however, through the choice of images, which are strikingly similar to those in *Time* magazine. It occurs rather through its layout, playing on strong visual symbols and catchy headlines directly referring to other disasters as well as to biblical scenes and iconology, with the apocalypse in the background.[44] Unsurprisingly, the main references are to popular knowledge of the Civil War and the end of the South ("Autant en emporte le cyclone" ["Gone with the Hurricane"], the burning of Atlanta),[45] the trope of war ("the airport is turned into a field hospital and some have already boarded for the other world"), classical and religious allusions (Breughel's *The Blind* and the biblical tale of the blind leading the blind [see fig. 3.3]), as well as other recent disasters ("Mississippi

Fig. 3.3. "In the heavy rain, some join together to survive." Photographer: Alvaro Canovas, *Paris Match.* Used by permission of *Paris Match*/Scoop.

Fig. 3.4. "America at half mast." Photographer: Agence Sipa. Used by permission of *Paris Match*/Scoop.

Tsunami") and national imagery with "America at half mast" collocating an old black woman and the American flag (see fig. 3.4). Clearly for *Paris Match,* America is sinking. To make its point, the magazine published a picture by Ben Sklar of an SUV being used as high ground by a group of black and white people, men, women, and children. The children wear life vests (the sign that they have received some help already, or perhaps own a boat). The composition clearly suggests a microcosm of (American) society, a sort of Noah's Ark under flood rains (although we learn from the caption that the white people are the rescuers helping out a black family). But, while *Paris Match* chose to see the sinking of the first-world power, *Newsweek,* which used the same image, titled it "High Water Heroics" and told the readers that "this family was saved."[46]

The most impressive difference between French and American news magazines, however, is how they deal with the law-and-order approach, clearly indicted by *Paris Match* with the image of a state-police armored vehicle with several heavily armed policemen oblivious to two black women asking for help (see fig. 3.5): "the only response: the army" and "they expected food, they got tanks," or with a scene of arrest titled: "military before humanitarian."[47] Clearly, for *Paris Match* the United States is but another corrupt third-world state, or even more to the point, it's treating Louisiana as another Afghanistan or Iraq.[48]

A Temporary Conclusion

Photographs published in the immediate aftermath of the hurricane and the ensuing flood are clearly marked by their "classicism." When dealing with catastrophes, photographers and picture editors tend to rely on a small number of identifiable rhetorical figures that have defined this type of reporting since the 1930s.

Fig. 3.5. "They were hoping for doctors and food. What they got was tanks." Photographer: Agence Sipa. Used by permission of *Paris Match*/Scoop.

One feature, however, makes Katrina's coverage stand out. Whatever the editorial stance of the publication (liberal, conservative) and its positioning (local, national, international), the photographs of Katrina's destruction display a high level of ambiguity, revealing the complexity of the actual situation of the city both in its history and after this latest major disaster. The ambiguity exists between an apparent return to the (frontier) origin of the nation, mythologically transforming the perennial climatic ordeal of the South into a perennial rebirth and rejuvenation, and the image of a city whose population appeared as fundamentally alien. If such a duality of vision does not explain the reality behind the image—and this image is far from monolithic—it is one more piece of evidence of the slow evolution of *cultural* features as well as of *visual* culture.

Notes

1. Unless otherwise specified, all websites cited were accessed on January 25, 2014.

2. Kendall L. Walton, "Transparent Pictures: On the Nature of Photographic Realism," *Critical Inquiry* 11, no. 2 (December 1984): 246–77.

3. See Jean Kempf, "L'Œuvre photographique de la 'Farm Security Administration': Quelques questions de rapport entre photographie et société," PhD diss., Université Lumière–Lyon 2, 1988, theses.univ-lyon2.fr/documents/lyon2/1988/kempf_j; Maren Stange, *Symbols of Ideal Life: Social Documentary Photography in America, 1890–1950* (Cambridge, U.K.: Cambridge University Press, 1989); Michele L. Landis, "Fate, Responsibility, and 'Natural' Disaster Relief: Narrating the American Welfare State," *Law & Society Review* 33, no. 2 (1999): 257–318.

4. See digitaljournalist.org/issue0512/jackson_intro.html. Two documents help us understand the conditions in which the photographers worked in New Orleans on the first days of the hurricane and when the levees broke: Marianne Fulton, "UT Katrina Coverage, Digital Photographer," with statements by Sloan Breeden, Mark Mulligan, Anne Drabicky, Meg Loucks, Rob Strong, and Ben Sklar, digitaljournalist.org/issue0510/dis_reed.html; and Eli Reed, "Photojournalism Students Cover Hurricane Katrina in Their First Leap into a Real-World Crisis," www.nieman.harvard.edu/reportsitem.aspx?id=100611. Those are ex post facto testimonies of seasoned professionals and young aspiring journalists. Necessary caution must be exercised towards those heavily edited testimonies, collected by fellow journalists and to some extent participating in an "operation of justification." The very collection and exhibition of the testimonies of the photographers themselves, however, is significant enough to make us realize the import of the act of taking the pictures in such situation, at least of the *mental representations* of the act. However, my conclusions as to the specific relation of the local professional photographers to their subject were confirmed by personal interviews conducted with one of the photographers, John McCusker (Lyon, March 30–31, 2011). These are congruent with other testimonies of war and documentary photographers and evidence the existence of a common shaping force in dramatic events for journalistic wit-

nesses, and the existence of a common professional culture among them. The quote is from digitaljournalist.org/issue0512/jackson_intro.html.

5. See David Eggers's excellent narrative *Zeitoun* (New York: Vintage Books, 2009) for similar accounts of the first-day rescues.

6. Bill Pierce, digitaljournalist.org/issue0509/pierce.html.

7. A U.S. Army general declared that the hurricane was "like a well planned attack" (*Times-Picayune*, September 14, 2005, 10) This phrase points at the running simile which undoubtedly served as justification for the use of the military in defense and not simply humanitarian duty. It also made the photographers not simply "civilian reporters" but let them compare themselves to the highest class of journalist in the romantic (and prestige) scale, the war correspondent. See www.msnbc.msn.com/id/14484343/ns/news-picture_stories/displaymode/1107/s/2/, image #4 and attached soundtrack. For "adventure" see, for instance, Ted Jackson: "the nutty part of this business [journalism] [is that] when everybody is getting out, you're running in" (part 3); Bill Pierce, digitaljournalist.org/issue0509/pierce.html; Donald Winslow in *Digital Photographer*, digitaljournalist.org/issue0511/winslow.html.

8. See digitaljournalist.org/issue0512/jackson_intro.html. Ted Jackson also explains in his video interview his line of conduct as to using his boat for rescue, a moral decision that he compares to the sinking of the *Titanic* and the use of life boats.

9. See Miles Orvell, *The Real Thing: Imitation and Authenticity in American Culture, 1880–1940* (Chapel Hill: University of North Carolina Press, 1989).

10. The current dissemination of images through the internet is not taken into account here as it seems that the images published in the traditional media still by and large shape the portrayal of the event.

11. See www.nieman.harvard.edu/reportsitem.aspx?id=100611, digitaljournalist.org/issue0510 /video-utkatrina.html, and www.bensklar.com/katrina.html.

12. For amateurs' shots, see Flickr's Katrina gallery, www.flickr.com/groups/45871688@N00 /pool/. These will not be discussed here but were used as a sort of test against which to look at professional images.

13. Russell Lee, "The hands of Mrs. Andrew Ostermeyer, wife of a homesteader, Woodbury County, Iowa," 1936, Library of Congress; Walker Evans, "Floyd Burroughs' Working Shoes," Alabama, 1936, published in *Let Us Now Praise Famous Men* (Boston: Houghton Mifflin, 1941).

14. See www.msnbc.msn.com/id/14484343/ns/news-picture_stories/displaymode/1107/s/2/, image #5.

15. See, for instance, a photograph by Chris Granger, *Times-Picayune*, September 15, 2005, A9, www.nola.com/katrina/pages/091505/A09.pdf, or an earlier one by Eliot Kamenitz, *Times-Picayune*, September 8, 2005, 13, www.nola.com/katrina/pages/090805/13.pdf.

16. For the latter, see the famous Dorothea Lange, *Migrant Mother*, or Arthur Rothstein, *Human Erosion*, or even Russell Lee, "An organ deposited by the flood on a farm near Mount Vernon," Indiana, 1937 (Library of Congress).

17. I am dealing here only with the "low" aerial shots made from helicopters, not the high-altitude shots made by NOAA (National Oceanic and Atmospheric Administration) (visible at www .katrinadestruction.com/images/v/noaa_overhead_gulf_coast_aerial_images/), one of which was published in the *Times-Picayune* of September 1, nor with NASA space shots of the hurricane.

18. See www.katrinadestruction.com/images/v/new+orleans+flood/.

19. *Times-Picayune,* September 2, 2005, 13; *Newsweek,* September 12, 2005, 18–19; *Paris Match* wrote: "New Orleans is sinking" (September 8–14, 2005).

20. See www.katrinadestruction.com/images/v/hurricane/swamped+school+buses.html and www.katrinadestruction.com/images/v/hurricane/15021w.jpg.html.

21. See also the shoe (*Times-Picayune,* September 11, 2005, 2), the Mardi Gras mask (*Times-Picayune,* September 10, 2005, 9), and the old people's home (*Times-Picayune,* September 15, 2005, 9).

22. For *Times-Picayune* images of return, see issues of September 6, 2005, "Coming Home" on the front cover, 11, 13; "Broken Dreams," 14, 17; "Picking Up, Cleaning Up," 17; "Facing the Aftermath," September 18 (p. 20), 19 (p. 15), but also the September 15 cover showing a woman and her dog amid her scant possessions with the headline "The Bill Comes Due," referencing insurance problems. In its September 1 issue, the *Times-Picayune* writes "Homeowners get first glimpse of devastation on North Shore" (13), showing a picture of a tent and three refugees in the foreground (seen from behind they are visually anonymous, although they are identified in the caption) and the scene of devastation with two houses perfectly intact.

23. *National Geographic,* August 2005, 77.

24. Ibid., August 2006, 42–43.

25. Alan Sekula, "On the Invention of Photographic Meaning," in *Thinking Photography,* ed. Victor Burgin (London: Macmillan, 1982), 106–7.

26. See Kempf, "L'Œuvre photographique de la 'Farm Security Administration'"; Stange, *Symbols of Ideal Life*; and Vicki Goldberg, *The Power of Photography: How Photographs Changed Our Lives* (New York: Abbeville Press, 1993), 135 ff.

27. *Time,* September 12, 2005. The same titles can be found in news coverage of almost all catastrophes, such as the Haitian earthquake, the Indonesian tsunami, and so forth.

28. This image in many ways resembles other famous war documentary pictures. See a variation on the theme with the article "Air Raid Victim" (*Life* cover, September 23, 1940).

29. The *Times-Picayune* rarely uses closeups or individuals for its front page, and here joins picture and sound with a semiotically open image: is this woman dancing? singing? crying? shouting? See www.nola.com/katrina/pages/090205/a01.pdf.

30. See also the rescue scene in *Times-Picayune,* August 31, 2005, 2.

31. *Times-Picayune,* September 2, 2005, 11, September 5, 2005, 4, September 8, 2005, 2, "Civic Minded Gunmen Patrol Algiers." The *Times-Picayune* September 10, 2005, cover for instance looks like a Vietnam or *M*A*S*H* scene.

32. Photo by Alex Brandon, *Times-Picayune,* September 8, 2005, 1, www.nola.com/katrina/pages/090805/1.pdf.

33. *Time,* cover of September 24, 2001, and 46–47.

34. See www.nola.com/katrina/pages/091305/01.pdf.

35. *Time,* September 24, 2001.

36. This picture is part of a set on the highly media-oriented presidential visit to New Orleans. Most of it is a series of carefully crafted "photo ops." The AP picture described above was published in a black-and-white version in the *Times-Picayune,* September 16, 2005, A3 www.nola.com/katrina/pages/091605/A03.pdf.

37. See www.magnumphotos.com/C.aspx?VP3=CMS3&VF=SearchDetailPopupPage&VBID=2K1HZS6YSTU9F&PN=34&IID=2K7O3R1U3NOP.

38. A pro-life group, Columbia Christians for Life, seems to have issued a press release comparing the shape of the hurricane seen from space to that of an unborn fetus. See www.dailykos .com/story/2005/8/31/0836/62623. I could not find any trace of the original claim on this group's website. The information seems to have been sent by e-mail, and as I am not aware of any public disclaimer, I assume this piece of information is reliable. Mayor Nagin also implied that Katrina was a "payback from God" for the invasion of Iraq and for black infighting (www.washingtonpost .com/wp-dyn/content/article/2006/01/16/AR2006011600925.html). On the negative image of the city, see Randy Sparks, "American Sodom: New Orleans Faces Its Critics and an Uncertain Future," *Nuevo Mundo Mundos Nuevos, Coloquios,* 2007, put online on May 8, 2007, nuevomundo.revues .org/3941, and his contribution in the present volume.

39. See www.nola.com/katrina/pages/090705/1.pdf.

40. *Times-Picayune* photographer Ted Jackson says it in so many words: "we lived in a frontier town now," and concludes, "hard times call for strong people" (part 4 of his interview).

41. Cover of September 19, 2005, issue.

42. See *Newsweek*'s review of the foreign press: "The Wonderful World of Oz" (September 19, 2005, 17).

43. *Time,* October 3, 2005. I will not deal here in detail with the use of information graphics and other visual representations including scientific images (usually of weather conditions), which are now commonly used by the media to complement their pictures and text. They carry all the authority of science—power but also mystery and distance from the object. One good example is the reduplication of a before-and-after (Hurricane Ivan) picture set by a totally redundant computer analysis of the texture of the earth's surface, whose function is less to explain than to assert the idea of a scientific analysis of those phenomena (what Roland Barthes would have called a sign of "science-ness"). See *National Geographic,* August 2005.

44. *L'Express* (September 5, 2005) did the same with a photograph reminiscent of those of 1930s Dust Bowl farmers, titled "The Grapes of Misery."

45. *Paris Match,* September 8–14, 2005.

46. Ibid.; *Newsweek,* September 12, 2005, 22.

47. Of course coverage by the *Times-Picayune* is much more varied as the local paper published many more images over the period.

48. A predictable incident took place around two captions made by different agencies, one of a young black man carrying a bag and described as "looting," and that of a white couple carrying bags and described as "finding food" (www.nytimes.com/2005/09/05/business/05caption .html?ex=1283572800&en=340529d0729a0b63&ei=5090). Although the debate was inconclusive, the question of the characterization of actions in those emergency situations was paramount as *Paris Match* asked: "looting begins, out of necessity or in order to steal?" displaying the image of two ordinary black women being held at gunpoint by the police like dangerous criminals (one of them opens her arms in complete abandonment) in a scene reminiscent of many TV series but also showing the absurd brutality of police action. Similarly one may note that one of the black persons in the image is described as "a Jazz musician, member of a gang."

"Wilt Thou Judge the Bloody City? Yea, Thou Shalt Show Her All Her Abominations"

Hurricane Katrina as a Providential Catastrophe

JAMES BOYDEN

Cataclysmic events, however physically tangible their consequences, seem always to provoke metaphysical explanations. The urge to make sense of disaster, akin to other categorizing impulses in human thought, reflects the need to order and organize, the better to understand the disorderly phenomena of life. Throughout history, people have struggled to make sense of catastrophic events and to find some meaning or purpose in the destruction and loss of life they cause. In the Christian West, cataclysm was often ascribed to God's wrath—the prototype here is the biblical story of the cities of the plain, Sodom and Gomorrah—but such events also consistently prompted grave religious doubts. How can a just God perpetrate or permit the destruction of innocents? Abraham remonstrated with the Lord over Sodom: "Wilt thou also destroy the righteous with the wicked?"[1] And perhaps most famously, believers and skeptics across the European world debated the implications of the devastating Lisbon earthquake of 1755.

Intellectuals, at least, habitually assume that the skeptics got the better of this argument, finding Voltaire's taunts irrefutable:

> Say ye, o'er that yet quivering mass of flesh:
> "God is avenged: the wage of sin is death"?
> What crime, what sin, had those young hearts conceived
> That lie, bleeding and torn, on mother's breast?
> Did fallen Lisbon deeper drink of vice
> Than London, Paris, or sunlit Madrid?[2]

And Voltaire was no kinder to those who posited that the meanings of terrestrial events must be parsed for beneficial consequences:

"All's well," ye say, "and all is necessary."
Think ye this universe had been the worse
Without this hellish gulf in Portugal?[3]

We may tend to assume a broad contemporary consensus in the post-Enlightenment West that calamities like earthquakes and storms are in fact natural events whose explanation falls within the realm of science. By definition, at least for rationalists, such disastrous occurrences lack supernatural causation, and they need encompass no intelligible meaning nor require metaphysical analysis, let alone offer rays of hope for the future. This is the position adopted, for instance, by Susan Neiman, who asserts that "plagues and floods and earthquakes ravage them all [the just and unjust] without distinction or warning. Before the Enlightenment, they were all known as natural evils; now we call them disasters, thereby recording our belief that nature has no moral categories."[4]

This seems eminently reasonable, but we ought to bear in mind that even the high tide of Enlightenment did not sweep away all contrary opinion. Musing upon the calamitous Peruvian earthquakes of 1746 and their ominous import for Christians in the Old World, the lord bishop of London, Thomas Sherlock, scoffed at "Little Philosophers, who see a little, and but very little into natural Causes, [and] may think they see enough to account for what happens, without calling in the Aid and Assistance of a special Providence; not considering, that God who made all Things, never put any Thing out of his own Power, but has all Nature under Command to serve his Purposes in the Government of the World."[5] The bishop had no qualms about ascribing moral categories to nature, concluding that "the dissolute Wickedness of the Age, is a more dreadful Sign and Prognostication of Divine Anger, than even the Trembling of the Earth under us."[6]

In a striking proof of the persistence of this viewpoint, the devastation of New Orleans by Hurricane Katrina in 2005 provoked a broad array of providential interpretations. Fundamentalist Christians immediately discerned the hand of God in the storm. The television evangelist Pat Robertson was quick off the mark. As the hurricane moved across the gulf, he delivered what in football would be termed a "lookout!" block: Katrina would hit New Orleans like a blitzing linebacker as "God's revenge for homosexuality." Clearly Robertson deduced that the deity was not to be deflected from this mission of righteous vengeance. While he claimed that his prayers had twice spared Virginia from hurricanes, this time the evangelist made no bid to still the wind and waves. He would at least assist the Lord in the mop-up operation, though; Robertson's charity, the so-called Operation Blessing, was high on FEMA's online list of organizations to which donations could be sent for storm relief.[7]

Robertson may have been first to see the awful judgment of the Almighty in Katrina's counterclockwise spin, and he was also the pioneer of what would be a popular explanation of God's motives on those terrible final days of August 2005. Divine loathing of homosexuality is of course a hardy perennial when it comes to religious explanations of cataclysm. (It must be said, though, that the precise failings of the original Sodomites remain surprisingly vague in the canonical version, Gen. 13:13: "But the men of Sodom were wicked and sinners before the Lord exceedingly.") Half a millennium ago, rabble-rousing preachers had already mastered the technique that Robertson employed to interpret Katrina. While he roiled the politics of Florence in the 1490s, the Dominican friar Girolamo Savonarola time and again insisted that homosexual behavior among the Florentines was the cause of the city's setbacks.[8] And in 1750, Bishop Sherlock was adamant on this score: "The unnatural Lewdness, of which we have heard so much of late, is something more than brutish, and can hardly be mentioned without offending chaste Ears, and yet cannot be passed over entirely in Silence, because of the particular Mark of Divine Vengeance set upon it in the Destruction of *Sodom* by Fire from Heaven. Dreadful Example!"[9]

Another early interpreter of New Orleans's cataclysm avoided the specificity of Pat Robertson's diagnosis. This was the Texas evangelist Kim Clement, who wrote well after the storm and the flooding, claiming to report a prophecy he had vouchsafed in a Houston church five and a half weeks before Katrina on July 22, 2005. (Regrettably, a retrospectively reported prediction cannot, in my view, wrest the palm of prophetic priority from Pat Robertson.) In any case, Clement said that he had foreseen that New Orleans would be destroyed for its unnamed "transgressions": "O New Orleans," his recollected prophecy continued, "God speaks to you from Houston tonight and says enough of this! For a judgment is coming says the Spirit of the Lord."[10]

In case flooded Louisianans were skeptical of divine pronouncements emanating from southeast Texas, similar messages also came from elsewhere. By August 31, Michael Marcavage, the leader of a Philadelphia group called Repent America, had managed to steal a more detailed glimpse of the divine design. For this pastor, it was no coincidence that the drownings of hundreds, the despair of thousands, the destruction of much of New Orleans, and the displacement of its residents had come "just days before 'Southern Decadence,' an annual homosexual celebration attracting tens of thousands of people" to New Orleans. Marcavage explained further that "'Southern Decadence' has a history of filling the French Quarters [*sic*] . . . with drunken homosexuals engaging in sex acts in the public streets and bars. . . . However, Hurricane Katrina has put an end to the annual celebration of sin. Although the loss of lives is deeply saddening, this act of God destroyed a wicked city."

Perhaps anticipating that some observers might wonder why the deity couldn't have stayed his wrath a few days so as to smite gay revelers alongside, say, bedridden nursing home residents, Marcavage made it clear that the Crescent City had habitually besmirched God's preferred sexual orientation as well. Not even heterosexual New Orleanians could get a pass, since the city "was also known for its Mardi Gras parties where thousands of drunken men would revel in the streets to exchange plastic jewelry for drunken women to expose their breasts and to engage in other sex acts. This annual event sparked the creation of the 'Girls Gone Wild' video series." (Here it must be admitted that giving birth to the "Girls Gone Wild" concept is not one of the Crescent City's most savory cultural achievements, though perhaps not so egregious a failing as to provoke a full-bore divine smiting.) Marcavage disagreed, however, concluding that, "From 'Girls Gone Wild' to 'Southern Decadence,' New Orleans was a city that had its doors wide open to the public celebration of sin. . . . Let us pray for those ravaged by this disaster. However, we must not forget that the citizens of New Orleans tolerated and welcomed the wickedness in their city for so long."[11] So, you see, there really were *no* innocent victims of Katrina in New Orleans.

Compounding the city's sexual sins, some Christian observers divined another cause for the Big Easy's cyclonic calamity. "They have devil worship," alleged Dwight McKissic, an Arlington, Texas, pastor, to a reporter for the *Austin American-Statesman*. He elaborated in a subsequent interview that "they openly practice voodoo and devil worship in New Orleans," adding that "you can't shake your fist in God's face 364 days a year and then ask, 'Where was God when Katrina struck?'" And Franklin Graham ("the hereditary evangelist," in James Gill's felicitous phrase) concurred that "there's been satanic worship, there's been sexual perversion" in the Crescent City. To be sure, neither of these evangelists was certain that Katrina amounted to an episode of divine chastisement. Graham declined to link the storm directly to the decadent doings on Bourbon Street. McKissic displayed unexpected thoughtfulness, musing that "sometimes God does not speak through natural phenomena. This may have nothing to do with God being offended by homo-sexuality. But possibly it does."[12] Whatever the Lord may or may not have been up to, though, Franklin and McKissic could agree that Katrina's consequences served Christian moral purposes. Graham asserted that "God is going to use that storm to bring revival," while Pastor McKissic saw substantial short-term benefits, asking, "Could it be God sent Katrina to purify the sins of America? Not just New Orleans. That week there was no gambling, no prostitution, no sins in New Orleans. It became one of the purest cities in America during that time."[13] And even a week of purity was surely a boon.

In South Carolina, the group Columbia Christians for Life found another reason to exult in the aftermath of the storm. A document on the CCL website trumpeted the joyful tidings that "There are no abortion centers open in New Orleans," and continued "It has been reported from the director of a Christian retreat center which is accomodating [*sic*] several hundred refugees, from the devastation in New Orleans and elsewhere, caused by the *Act of God* named Hurricane Katrina, that *the child-murder-by-abortion centers in New Orleans are all shut down!!!* The bloody city of *New Orleans* (Ezekiel 22:2–4) *had five operating childkilling centers. . . .* As sad as it is to see the heartbreaking loss of life and the suffering of people in New Orleans, we can only *give praise to God for sparing the lives of the innocent unborn who have been murdered by the tens of thousands in New Orleans and the rest of the state of Louisiana* year-after-year-after-year."[14]

The passage from Ezekiel cited by CCL opens with "Now, thou son of man, wilt thou judge, wilt thou judge the bloody city? yea, thou shalt show her all her abominations." Columbia Christians for Life shared none of the hesita-tions about the providential nature of the storm or the temporizing about divine motives that characterized some of their evangelical brethren. Their literature flatly asserted that "God brings his Just Judgment through natural disasters (*Acts of God,* not some goddess the media calls 'Mother Nature'!)" Somewhat oddly, scriptural authority on this score was buttressed by a CCL reference to the early American statesman George Mason, quoted to the effect that "As nations cannot be rewarded or punished in the next world, they must be in this. By an inevitable chain of causes and effects, providence punishes national sins by national calamities."[15] And CCL offered breathtaking proof that it was indeed abortion that had spurred God's wrath in the case of Ka-trina. In an email sent to various news outlets on August 30, 2005, the group proclaimed, "Hurricane Katrina satellite image looks like 6-week fetus." How so? The email explained that "the image of the hurricane with its eye already ashore at 12:32 p.m. Monday, August 29, looks like a fetus (unborn human baby) facing to the left (west) in the womb, in the early weeks of gestation (approx. 6 weeks). . . . Even the orange color of the image is reminiscent of a commonly used pro-life picture of early prenatal development."[16] With his vaunted attention to detail, the Almighty had not just numbered the hairs of our heads (Matt. 10:30) but evidently had also preordained that coloring conventions for meteorological satellite images would match those adopted by anti-abortion activists.

It is perhaps only fair to acknowledge that most of these lip-smacking reactions to New Orleans's devastation came from far away and at least nominally higher ground, in Philadelphia, or Columbia, South Carolina, or Virginia Beach. Sadly if not surprisingly, some distant observers felt the call to parse the woes of Louisianans and preach the facile, cruel, and frankly asinine moral lessons they derived from them.[17] But what can we make of Bill Shanks, a Crescent City resident, pastor of the New Covenant Fellowship of New Orleans? Having evacuated to Jackson, Mississippi, Shanks was interviewed on September 2, 2005, by reporters for a Christian news service, and expressed his satisfaction with the godly work performed by Hurricane Katrina. His remarks recapitulated the major themes sounded by his evangelical brethren: "New Orleans now is abortion free. New Orleans now is Mardi Gras free. New Orleans now is free of Southern Decadence and the sodomites, the witchcraft workers, false religion—it's free of all of those things now. . . . God simply, I believe, in His mercy purged all of that stuff out of there."[18]

This line of reasoning, however, was hardly universal among American evangelicals. Prominent preachers like Franklin Graham held themselves at least somewhat aloof from the notion of Katrina-as-godly-scourge. Bishop T. D. Jakes, leader of the Potter's House ministry in Dallas, sounded a positively Voltairean note in rejecting providential interpretations of the hurricane. "I'm not sure," he said, "that New Orleans is any more wicked than Paris or Los Angeles or New York or even Dallas. . . . There were children who died. Innocent babies who died. Aged nursing home patients who died. To say that God is judging New Orleans and that these people died and some felons escaped, I personally have a problem with it. . . . To point a finger in the faces of people who are burying folks and say, 'This is why they died,' I don't see that."

Bishop Jakes also challenged Christians on the ground of biblical interpretation. According to the *Dallas Morning News,* he observed that "many people miss a key point in the story of Sodom and Gomorrah. 'God always brought the righteous out before he destroyed the city.'"[19] In a parallel vein, Christian author David Haggith noted that God had traditionally found ways to warn his faithful of wrathful visitations to come. "This storm," he wrote, "has most likely ended the lives of hundreds of Christians in New Orleans who had no prophetic warning that they should flee. . . . What kind of a God is that, which some people are so willing to put forward, who is unable to communicate to one who wants to listen, so he has to kill many? Whatever happened to God's ability to raise prophets *before* the big event?"[20]

And Chuck Kelley of the New Orleans Baptist Seminary spoke of manifestations of divine mercy in Katrina distinct from the sanguinary gifts of God celebrated by Columbia Christians for Life and their ilk. Kelley told reporters: "Imagine what would have happened if [New Orleans] had taken a direct hit. The levee did not break until after the storm was clear and the winds had died down and the rescue workers were able to get out. . . . It's a terrible tragedy, and we still don't know the scope of it—but the evidences of God's mercy are there."[21] An alternative interpretation of God's mercy in the storm's aftermath came from a man identified only as Yovi, an Israeli commando-for-hire whose firm had been engaged by prominent New Orleanian Jimmy Reiss to guard the gated Audubon Place community uptown. Speaking to a reporter on September 11, 2005, the mercenary said, "I spoke to one of the other owners on the telephone. . . . I told him how the water had stopped just at the back gate. God watches out for the rich people, I guess."[22]

While American evangelical providentialists were focused on explaining God's targeting of the Crescent City—an Oregon preacher boiled it down, saying that New Orleans "is the epitome of a place where they mock God"[23]— similarly disposed commentators from other faiths and foreign nations tended to interpret Katrina as a judgment on the United States and its political leaders. From Iraq, a spokesman for Al-Qaeda in Mesopotamia and its chief Abu Musab al-Zarqawi opined that "God attacked America and the prayers of the oppressed were answered. . . . The wrath of the All-powerful fell upon the nation of oppressors."[24] Similar sentiments emanated from what one might have presumed were friendlier confines in the Middle East. From the Kuwaiti Ministry of Endowment, Muhammad Yusuf al-Mlaifi insisted that "it is almost certain that this is a wind of torment and evil that Allah has sent to this American empire."[25] Meanwhile Rabbi Ovadia Yosef of the conservative Israeli Shas party "said that Katrina was George Bush's punishment for supporting the Gaza pullout."[26]

Closer to home, in the weeks after the storm, Louis Farrakhan of the Nation of Islam told crowds in Philadelphia and Dallas "that the devastation caused by Hurricane Katrina was divine punishment for the violence America had in-flicted on Iraq," and that "the judgment of God now has entered America in a way that is going to get worse and worse and worse. . . . God is whipping Amer-ica, and he's whipping us." In Houston, Farrakhan suggested that God's dis-pleasure encompassed America's racism as well as its imperialism: "Maybe God ain't pleased. Maybe this caste system that pits us against each other has to be destroyed and something new and better put in its place."[27] In Austria, Father Gerhard Maria Wagner, then the Catholic rector of Windischgarsten, observed, "It is surely not an accident that all five of New Orleans' abortion clinics, as well as nightclubs, were destroyed." Presumably it still was no ac-cident even though Wagner's assertion is utterly fact-free. The priest was not, however, entirely negative on the subject of New Orleans, noting, "It's not just any old city, but the people's dream city with the best brothels and the most beautiful whores." Perhaps Wagner's views on the Crescent City, in tandem with his earlier attacks on Harry Potter novels, formed part of what qualified him to be elevated to the post of auxiliary bishop of Linz by Pope Benedict XVI in January 2009.[28]

And lest we think that only men of faith[29] pursued these lines of reason-ing, rest assured that more secular figures could parse the meteorological omens as well. Commenting on the storm in early September 2005, the Ger-man environment minister, Jürgen Trittin, for example, argued that George Bush "has closed his eyes to the economic and human damage that natural catastrophes such as Katrina—in other words, disasters caused by a lack of climate protection measures—can visit on his country."[30] (Remember that the Columbia Christians for Life were insistent that Katrina as act of God should not be ascribed to that pagan goddess "Mother Nature." Their thought pat-terns and Herr Trittin's are not so very far apart; for the German, environmen-tal sinners face the horrific visitations of Mother's wrath.) And in the United States, Robert F. Kennedy Jr. observed—mangling a biblical metaphor along the way—that "now we are all learning what it's like to reap the whirlwind of fossil fuel dependence. Our destructive addiction has given us a catastrophic war in the Middle East and now Katrina." Branching out from the typical focus on George W. Bush, Kennedy argued that Governor Haley Barbour of Missis-sippi should shoulder blame for Katrina on account of his "central role . . . derailing the Kyoto Protocol."[31]

The most widely publicized remarks along these lines came months later, in Mayor Ray Nagin's speech at New Orleans City Hall on the occasion of the Martin Luther King holiday in January 2006. Before his infamous peroration about "a chocolate New Orleans," the mayor mused that "surely God is mad at America, he's sending hurricane after hurricane after hurricane and it's destroying and putting stress on this country. Surely he's not approving of us being in Iraq under false pretense." Perhaps envisioning the deity as a multitasker, though, Nagin divined another godly admonition in the city's cataclysm: "But surely he's upset at black America, also. We're not taking care of ourselves. We're not taking care of our women. And we're not taking care of our children when you have a community where 70 percent of its children are being born to one parent."[32] Clearly Hizzoner had been doing some profound thinking in the months since the storm.

The MLK Day speech attracted a barrage of derision and opprobrium, mostly for its "chocolate city" tagline, but Nagin's binary interpretation of divine intentions revealed one of the traits that made him a fascinating, if ultimately baffling, politician. His providentialism was on the one hand perfectly calibrated to his constituency—among New Orleanians in those days, after all, who did not wish some avenging nemesis on the Bush Administration?—and on the other hand it could hardly have been better calculated to outrage his base of support. Was the mayor really implying that the utter destruction of homes and neighborhoods, massive loss of life, complete economic dislocation, and a painful, seemingly interminable diaspora should be read as God's wakeup call to African Americans? Could he possibly mean that Katrina pointed the way to a brighter tomorrow for black America and the renewed "chocolate city"? Well, yes, in fact—at the Wheeler Avenue Baptist Church in New Orleans in early March 2006, Mayor Nagin would assert, "Sometimes God has to do certain things to kind of clear the way for you to get to a better place."[33]

Nagin's speech stands at the hinge point where providentialist fantasies of vengeful destruction metamorphose into the Panglossian optimism that insists that all is for the best in this best of all possible worlds, or that God inflicts no crosses that we cannot bear, or as the hockey moms said in Sarah Palin's America, "when God gives you lemons, make lemonade." A fascinating project would be to assemble and analyze the catalog of redemptive fantasies and improving aspirations that were if anything more common than scenarios of divine wrath in the aftermath of the hurricane, and which continue several years on to obscure hard realities in New Orleans and beyond. But such a sequel will have to await another occasion. Perhaps a fitting conclusion, though, is that the continuing appeal of the providentialist line of irrationality stems from its duality: first from the devout longing of the impotent to see their *bêtes noires* struck down by what they imagine as cosmic justice, and second from the desperate hopes of decent people—and my two categories are by no means mutually exclusive—their insatiable craving to believe that suffering shall not have been in vain.

Notes

1. Gen. 18:23.

2. Voltaire (François Marie Arouet), "Poem on the Lisbon Disaster, or an Examination of the Axiom, 'All is Well,'" in *Toleration and Other Essays,* trans. Joseph McCabe (New York: G. P. Putnam's and Sons, 1912), 255.

3. Ibid., 256.

4. Susan Neiman, "Evil Acts, Evildoers: Moral Clarity after 9/11," *Christian Century,* May 20, 2008, 29.

5. Thomas Sherlock, *A Letter from the Lord Bishop of London, to the Clergy and People of London and Westminster; On Occasion of the Late Earthquakes* (London: Printed for John Whiston in Fleetstreet, 1750). Available online at Project Gutenberg, www.gutenberg.org/cache/epub/26204/pg26204.txt.

6. Ibid. Also quoted in Alessa Johns's introduction to *Dreadful Visitations: Confronting Natural Catastrophe in the Age of Enlightenment,* ed. Alessa Johns (New York: Routledge, 1999), xxiii, note 5.

7. Tom Feran, "This 'Blessing' Is Mysterious Indeed," *Cleveland Plain Dealer,* September 6, 2005, E1.

8. See, for example, Richard Trexler, *Public Life in Renaissance Florence* (Ithaca, N.Y.: Cornell University Press, 1991), 350.

9. Sherlock, *A Letter from the Lord Bishop.*

10. "July 22, 2005—Houston, TX," *The Sound of Kim Clement,* www.kimclement.com/words/2005/July222005.htm (accessed October 2, 2008); Selwyn Crawford, "In Disasters, Some See the Wrath of God," *Dallas Morning News,* October 5, 2005.

11. "The Real Cause of Hurricane Katrina?" *Salon.com,* August 31, 2005, dir.salon.com/story/politics/war_room/2005/08/31/cause/print.html (accessed September 30, 2008); Mel Seesholtz, "Katrina and Religion," *Counterbias: Online Opinion Armed with the Truth,* September 7, 2005, www.counterbias.com/399.html (accessed October 1, 2008); Richard Roeper, "God's Reasoning for Katrina Far Too Complex for Us," *Chicago Sun-Times,* September 13, 2005, 11.

12. Gardner Selby, "Perry Mum on Pastor's Musings about Katrina Purifying," *Austin American-Statesman,* September 11, 2005, B3; Crawford, "In Disasters, Some See the Wrath of God"; James Gill, "Earthly Critics Take Up Where God Left Off," *New Orleans Times-Picayune,* October 19, 2005, 7.

13. Gill, "Earthly Critics"; R. G. Ratcliffe, "Ministers Ask Faithful to Help Ban Gay Marriage," *Houston Chronicle,* September 18, 2005, B1.

14. Columbia Christians for Life, www.christianlifeandliberty.net/news3.htm (accessed October 2, 2008); the quoted document, in Microsoft Word format, may be downloaded from the site.

15. Ibid.

16. Ibid.; "Is Katrina God's Punishment for Abortion?" *Salon.com,* August 30, 2005, dir.salon .com/story/politics/war_room/2005/08/30/hurricane/print.html (accessed 7 Oct. 2008). *Salon* also reproduces the supposedly comparable satellite and fetal images.

17. For an eloquent colloquial reflection on similar behavior, listen to Steve Goodman's song "Somebody Else's Troubles," recorded on Buddah Records BDS 5121, *Somebody Else's Troubles* (1973).

18. Jody Brown and Allie Martin, "New Orleans Residents: God's Mercy Evident in Katrina's Wake," *AgapePress,* September 2, 2005, www.wench.org/forums/archive/index.php/t-9540.html, citing headlines/agapepress.org/archive/9/22005b.asp (the latter now a dead link). Shanks is similarly quoted in the *Times-Picayune,* September 8, 2005, A15, and by Richard Roeper in the *Chicago Sun-Times,* September 13, 2005, 11.

19. Crawford, "In Disasters, Some See the Wrath of God."

20. Quoted in Valerie Bridgeman Davis, "Retribution as First Response: Did God Punish New Orleans?" in *The Sky Is Crying: Race, Class, and Natural Disaster,* ed. Cheryl A. Kirk-Duggan (Nashville: Abingdon Press, 2006), 9. Like many others, Haggith had perhaps been deaf to Kim Clement's Houston prophecies.

21. Quoted in Brown and Martin, "New Orleans Residents: God's Mercy Evident in Katrina's Wake."

22. Jamie Wilson, "Mercenaries Guard Homes of the Rich in New Orleans," *The Guardian* (London), September 12, 2005, International 22.

23. David Crowe of Restore America, quoted by Crawford, "In Disasters, Some See the Wrath of God."

24. Seesholtz, "Katrina and Religion."

25. Tony Parkinson, "It Is Ludicrous to Blame the President for All the World's Woes," *The Age* (Melbourne, Australia), September 6, 2005, 13. Sadly, this line of argument recurred in the aftermath of Hurricane Sandy's devastation in the Caribbean and the northeastern United States in autumn 2012; see Kevin Rawlinson, "Muslim Clerics Say Superstorm Sandy Is God's Punishment for Film That Mocked Prophet Mohammad," *The Independent* (London), November 2, 2012, www .independent.co.uk/news/world/americas/muslim-clerics-say-superstorm-sandy-is-gods-punish-ment-for-film-that-mocked-prophet-mohammad-8276747.html (accessed November 2, 2012).

26. Marjorie Ingall, "Vengeance Is Mine, Sayeth Everybody," *The Forward* (New York), October 7, 2005, 16.

27. Seesholtz, "Katrina and Religion"; Crawford, "In Disasters, Some See the Wrath of God."

28. "Pope Names Ultra-conservative in Linz," Agence France-Presse, January 31, 2009, at services .inquirer.net/print/print.php?article_id=20090131-186711 (accessed January 31, 2009); James Gill, "Holy Nutcase!" *Times-Picayune,* February 4, 2009.

29. It is striking that in the reports I could find there is no evidence of women making the argument that Katrina represented an act of divine wrath.

30. Edward Rothstein, "Seeking Justice, of Gods or the Politicians," *New York Times,* September 8, 2005, E1; Parkinson, "It Is Ludicrous to Blame the President."

31. Parkinson, "It Is Ludicrous to Blame the President"; Joe Klein, "Listen to What Katrina Is Saying," *Time,* September 4, 2005, www.time.com/time/printout/0,8816,1101282,00.html (accessed September 30, 2008).

32. Transcript in *Times-Picayune,* January 17, 2006.

33. James Gill, "The Elect and the Elected," *Times-Picayune,* March 10, 2006, Metro 7.

Naturalizing Disaster

Neoliberalism, Cultural Racism, and Depoliticization in the Era of Katrina

ANDREW DIAMOND

Injustice and Quiescence

It was around September 2, 2005, four days after the winds of Hurricane Katrina had created a storm surge that topped the levee system and left some 80
percent of the city under water, that what had begun as a "natural disaster" or
"an act of God" began to seem more like something political. It was hardly surprising that, as images of flood-ravaged New Orleans flowed through television screens across the nation, most people understood what had befallen the
city as a *"natural* disaster." Yet, another story began to take shape as images
of poor black folks stranded on rooftops began to intermingle with questions
and allegations about FEMA's impossibly slow relief response and President
Bush's astonishing lack of concern. On September 2, expressions of outrage
surfaced across much of the media landscape. As Americans sipped their
morning coffee, they saw CNN anchor Soledad O'Brien blast FEMA's hapless
director Michael Brown for not having executed a massive airdrop of food after four days, when in Banda Aceh, Indonesia, she insisted, tsunami victims
were being fed after two. They heard Mayor Ray Nagin's expletive-studded
tirade against the federal government a day earlier again and again. More and
more airtime was given to interviews with incensed black leaders, many of
whom began to connect the dots between the inept rescue operation and the
country's continuing legacy of racial discrimination. And rapper Kanye West
used his appearance on NBC's *Concert for Hurricane Relief* that night to tell
a live national audience that "George Bush doesn't care about black people."[1]

   The visibility of such expressions of racially charged protest on the national stage signified the development of a media-induced "injustice frame"—
a term used by William Gamson to describe the cognitive frameworks whose

capacity to generate and channel "the righteous anger that puts fire in the belly and iron in the soul" are critical to collective social protest.[2] While Gamson and others engaged in framing analysis are mainly concerned with the ways in which movement leaders and institutions work to construct "collective action frames"—often in opposition to those put in place by the mass media—the injustice frame that was beginning to take shape around the circumstances of the government's failed relief effort was largely a by-product of the media coverage itself.[3] And the promotion of an oppositional frame centered on a sense of injustice that appeared to be racial in nature represented a potential paradigm shift for the political culture of "postracial," "post–civil rights" America, where appeals to bipartisanship, patriotism, and color-blindness have reduced the move to address racial issues to "playing the race card," engaging in "special interest" politics, or acting "politically correct." In the Katrina-era United States, it bears pointing out, Martin Luther King's famous dream of a nation in which his children would "not be judged by the color of their skin but by the content of their character" has been recuperated by conservatives to justify rolling back race-conscious policies aimed at closing the widening gap between blacks and whites in education and employment.

This is not to say that this racial injustice frame suddenly became hegemonic. In fact, in this confused moment, as Americans sought ways to make sense of the spectacle before them, a number of cognitive frames jostled for position. Indicative of this interpretive scrum was a particularly emotion-filled broadcast of CNN's *Lou Dobbs Tonight,* in which several competing perspectives were given relatively equal exposure. There were expressions of indignation at the behavior of alleged looters; calls for holding the federal, state, and local officials accountable; sound bites from black leaders alleging racism; and an interview with North Carolina Representative Melvin Watt, the chair of the Congressional Black Caucus, claiming that what was transpiring was "a class situation more than a race situation."[4]

Such divergent reactions revealed what Robert C. Lieberman has called a "fundamental dualism in U.S. political culture, its ambivalent embrace of both color blindness and race consciousness."[5] Indeed, a tug-of-war between these two positions was surely at play in much of the media coverage, and, for the first time in perhaps decades, race consciousness seemed ready to regain its former prominence in the public sphere. Indeed, as several recent studies have shown, color-blindness was at the ideological heart of grassroots movements against busing, affirmative action, and welfare programs that propelled the conservative ascendency in the 1970s and 1980s.[6] Not surprisingly, in its own attempt to spin the meaning of the event, the Bush administration drew from this tradition. "The storm didn't discriminate and neither will the recovery effort," the president declared. And Condoleezza Rice, the federal government's highest-ranking person of color, spoke of "Americans . . . pulling together to help Americans." Yet, so great was the surge of race consciousness over the levees of color-blindness and into the national discussion that, in his first major address to the nation from New Orleans, President Bush performed a dramatic about-face on his earlier color-blind position, issuing instead an unequivocally racial interpretation of the situation. "As all of us saw on television," the president declared, "there's also some deep, persistent poverty in this region as well." "That poverty has roots in a history of racial discrimination, which cut off generations from the opportunity of America."[7]

Since the racial injustice frame opened up an enormous opportunity for black protest politics, why was this opportunity ultimately missed? My purpose is not to argue for a racial over a class interpretation of what transpired in New Orleans after the hurricane struck—to take issue, for example, with Adolph Reed's just assertion that class was a better "predictor" of who left the city and who was left behind.[8] Rather, it is to try to understand how an event so seemingly loaded with provocative racial meaning could fail to elicit a meaningful political reaction. Even if the racism on display in the response to the situation in New Orleans was heavily class-infected, as Reed and others have argued, racism was spectacularly on display nonetheless. If the racial implications of the slow response of the federal government and the callous comments made by federal officials were matters for debate, much less ambiguous was the marked tendency of the media and the American public to view black flood victims as looters and "refugees"—circumstances that a number of black leaders were quick to point out. Then there was the incident at the border of Gretna, where the city's policemen fired shotguns over the heads of hundreds of African Americans to prevent them from crossing the Crescent City Connection Bridge and seeking refuge there.[9] And yet, such overt evidence of the lingering injuries of race, which even the president himself duly acknowledged, failed to muster any kind of sustained and nationally visible protest mobilization.

The project of analyzing quiescence is much trickier business than analyzing mobilization, since it is always difficult to begin with the assumption that actions or events should have or could have happened. The task becomes even more complex, moreover, when one assumes that those actions or events should have taken the form of an organized political movement. Here we must consider not only the ideological terrain, where perceptions of injustice and the need for action crystallize (or not), but also the organizational architecture in place for structuring protest actions. This is a project very different from the one recently undertaken by Michael Katz in his attempt to explain why American cities, with a few exceptions, have largely escaped outbreaks of civil violence since the early 1970s. In fact, in justifying this kind of inquiry, Katz points out that, while protestors in France took to the streets in a wave of collective civil violence after the death of two youths at the hands of police in suburban Paris, "[e]ven the botched response to Hurricane Katrina did not provoke civil violence" in the United States.[10] Ironically, the acts Katz labels as "civil violence"—"burning, looting, sniping at police"—all figured prominently in the media coverage of Hurricane Katrina, but we later learned that such circumstances were wildly overstated and often falsely reported. In fact, widespread civil violence never threatened to break out, neither in New Orleans nor anywhere else. That it did not might have something to do with the investment of many African Americans in the idea that legitimate forms of protest would surely prevail. Indeed the Los Angeles riot of 1992 began not after the airing of the video of motorist Rodney King's horrific beating, but rather, much later, after the brutality went unpunished by the judicial system.

To be sure, in earlier moments, the debacle in New Orleans might have sparked explosions of the kind witnessed in France several years ago—most certainly this would have been the case in the 1960s, and perhaps also in the early 1990s, when unrest swept through Los Angeles, Las Vegas, Atlanta, and several other cities after the acquittal of the police officers who had beaten King. Of course it can be argued that the miscarriage of justice in New Orleans was, while on a much greater scale, not nearly as clear-cut as that which characterized the Rodney King affair. And yet, large majorities of African Americans polled just after the flood did indeed view the meaning of Katrina in starkly racial terms. For example, a Pew survey after the flood revealed that 90 percent of blacks thought the disaster showed that racial inequality remains a major problem, and 84 percent believed the government would have responded more rapidly had the victims been white.[11] The perception of injustice was thus widespread, and it was an injustice, we should remember, that cost thousands of lives and put many thousands more at grave risk. Nonetheless, it comes as no surprise that the situation did not provoke collective civil violence. With 80 percent of New Orleans underwater and its inhabitants either evacuated or fighting for survival, local collective civil violence was never a possibility, and for the rest of the nation, there was never a single decisive incident that could have provided a spark. Harder to explain, given the kind of national discussion that was underway during the first few weeks of September, however, was the lack of an organized protest mobilization in the weeks after the flood. By the mid-term elections in November, just over two months later, few black leaders were talking about the issue of race in New Orleans, and the mainstream media outlets were treating those who *were* like troublemakers.

Why then didn't the sense of racial injustice that surfaced in early September gain traction within black political and civic institutions and build into something resembling a protest movement? In an era in which, with the notable exception of the powerful movement for "immigration justice," national, broad-based mobilization has largely been the domain of right politics, it is tempting to attribute this quiescence to the organizational inadequacy of black institutions. Yet, notwithstanding the claims of cultural critics like Robert Putnam that the United States has seen a dramatic decrease in "social capital" and "social networks" since the 1960s—the cultural bonds that make collective action and civic engagement possible—it would be difficult to argue that African American communities do not possess the networks and mobilizing structures to facilitate powerful forms of social protest against perceived injustices.[12] The National Association for the Advancement of Colored People (NAACP) and the National Urban League (NUL) are but two among several examples of well-funded black political institutions with extensive networks stretching into communities across the country. Even more important, however, are the numerous faith-based and campus-based organizations throughout the country, whose mobilizing potential was revealed in 1995, when some 800,000 converged on Washington for the Million Man March.[13]

A similar kind of mobilization to address the racial injustices exposed in New Orleans was thus organizationally feasible. Understanding why it did not occur requires examining why the ideological terrain provided such shaky and ultimately unstable foundations for the racial injustice frame that had materialized in early September of 2005. How did the ideological context surrounding Hurricane Katrina work to depoliticize the racial inequalities that were so vividly revealed by the media coverage of flooded New Orleans? Specifically, two interrelated rationalities were vital to this process of depoliticization—neoliberalism and cultural racism.

Considering neoliberalism as a rationality (rather than as an ideology) highlights its nature as a naturalized form of political reason that, like cultural racism, has the capacity to structure the meaning of social and political phenomena. While most discussions of neoliberalism as a policy approach involve its promotion of privatization schemes, its protection of the free flow of capital and goods, and its maximization of corporate profits through the elimination of taxes and regulations, this analysis rests on a conception of neoliberalism as a central strand of popular common sense that, since at least the 1980s, has steered Americans away from social, structural, and ultimately political interpretations of a range of issues.[14] But, to be able to accomplish this, neoliberalism has also required the assistance of cultural racism—a complementarity that played out powerfully in the days and months after Hurricane Katrina.

According to a number of polls and surveys, white civil society largely refused to view Katrina in racial terms.[15] For example, only 31 percent of white Americans polled by *Newsweek* magazine believed that race played a role in the government's slow response, and, according to the same Pew survey referred to above, only 38 percent felt that the disaster demonstrated that racial inequality remained a major problem. Michael Dawson has recently argued that these dramatically different perceptions of the racial dimensions of Katrina reveal that blacks and whites in the United States inhabit "racially separate publics and public spheres."[16] While there is some truth to this contention, the rationalities that eventually proved decisive in shaping the meaning of Katrina—both neoliberalism *and* cultural racism—crossed the black-white racial divide. However, this is not the only reason that an analysis of political quiescence in the Katrina era must consider the black response along with the white one. If black civil society appeared to embrace very different feelings regarding the Katrina disaster, this does not mean that its potential for generating effective political mobilization did not depend somewhat on what was transpiring within white civil society. As the successes of the civil rights movement of the 1950s and early 1960s demonstrate, the viability of broad-based black mobilization rests heavily on a sympathetic white public preparing the ground for its growth. Such was not at all the case in the aftermath of Katrina, when the white investment in a color-blind interpretation of the event served to radicalize black activists and leaders attempting to work within the racial injustice frame. The rejection of the racial perspective by white America was so strong that, by January of 2006, a *Boston Globe* columnist was referring to the idea that Katrina affected blacks more than whites as "racial paranoia."[17]

Neoliberal, Color-blind, and Culturally Racist

It seems almost absurd in the present moment to take seriously the idea that white civil society—even if the perception of a grave injustice visited upon black flood victims by the federal government had taken hold of it—could have produced meaningful acts of protest on behalf of African Americans. If one could argue that such biracial solidarity, at times, characterized the civil rights and antiwar protests of the 1960s, the end of this era witnessed the gradual disintegration of the organizational and ideological ties that had made such mobilizations possible. In the first half of the 1990s, a series of spectacular, national events seemed to define the unbridgeable gap between whites and blacks in the political sphere. In the Los Angeles riot of 1992, whites watched black youths savagely beating trucker Reginald Denny just because he was white and in the wrong place at the wrong time. In 1993, they gazed in horror as African Americans celebrated the acquittal of O. J. Simpson, a man they understood to be clearly guilty of brutally killing his white wife and her lover in cold blood. And, in 1994, they viewed a "million" men marching on Washington in what many took to be a powerful expression of racial separatism. Hence, by the time Katrina made landfall in 2005, five years into the reign of a president who had made history by becoming the first head of state not to address the NAACP, white civil society had become largely cut off from black progressive politics. That the Democratic Party made almost no effort to arouse indignation about the racial injuries of Katrina on the eve of the midterm elections is suggestive of this rupture. Its refusal to seize the occasion was due to there being, in reality, no occasion to seize. Polling data indicated that most whites were not receptive to such ideas, and without such support, Democratic officials—members of the Congressional Black Caucus and those on the party's left flank alike—feared being cast as engaging in divisive politics.

This is not to say that considerable numbers of white activists and aid volunteers did not take part in relief and reconstruction efforts in post-flood New Orleans. Thousands of whites, many of them college students, joined the AFL-CIO, ACORN, and a range of local groups like the Common Ground Collective to provide services and help rebuild the city. Relief, however, is not nearly the same as political protest—a situation captured by the Common Ground Collective's adoption of the motto "Solidarity, Not Charity."[18] Indeed, in the aftermath of Katrina, relief quickly became the dominant register of white engagement across the political spectrum, and as such, a key frame for understanding what had befallen those in need of help. On the one hand, the call to relief sustained the idea that what was being witnessed was a "natural" disaster, and, on the other, it deflected attention away from the state's abdication of its responsibility in this area. In another era, such cries to private organizations and citizens to pick up the slack for an inept government rescue operation that left tens of thousands in grave danger might have provoked angry recriminations focusing on government accountability; in the Katrina-era United States, where the state has increasingly devolved its responsibility for the welfare of its disadvantaged citizens to private charities and so-called private "faith-based initiatives," they barely raised a quiver of doubt.

The rise of faith-based initiatives during the presidency of George W. Bush is symptomatic of a political culture shaped by the convergence of neoliberalism and neoconservatism. This convergence has worked to accelerate processes of de-democratization that have come to produce, as Wendy Brown has argued, "an undemocratic citizen . . . who expects neither truth nor accountability in governance and state actions."[19] Indeed, the project of shifting the provision of emergency relief services and other forms of welfare support from the government to private charities is, to be sure, a by-product of a political culture held captive by neoliberal ideas of market efficiency and small government; the emphasis placed on "faith-based" forms of charity reveals the way in which such neoliberal forms of reasoning work together with neoconservative ones regarding the role of religion in public life. The result, as poignantly demonstrated in the aftermath of Katrina, is the replacement of political engagement by charitable activities, which, in this case, were defined by color-blind and sentimental slogans—"Americans helping Americans," for example—that tended to "whitewash" the issue of accountability and neutralize emotions of injustice. Indicative of just how far relief, as a form of civic engagement, took citizens away from progressive political sensibilities, was the appearance of Pat Robertson's Operation Blessing on the list of charity organizations promoted by FEMA. Those providing aid through this organization were thus validating a religious leader who had publicly interpreted the hurricane as God's punishment for the abortions performed in the country.

While such neoconservative ideas about divine intervention certainly played some role in shaping the meaning of what had befallen the Big Easy for devout evangelical Christians, most whites viewed the disaster through the lens of neoliberal-inflected common sense. If Roosevelt's New Deal administration of the 1930s led Americans to imagine a federal government that was accountable for the public welfare in a myriad of ways, such notions have largely ceased to exist in the United States of today. In the minds of many, the federal government's role now begins at the nation's borders and extends out into the world, where the real threats to national security hide and plot. Thus, white Americans placed remarkably little blame on the federal government for the deterioration of conditions in New Orleans after the flood, and expressed little concern that hurricane victims would receive sufficient government assistance. According to one survey, for example, only 42 percent of whites believed evacuees deserved a great deal of assistance, while 50 percent felt this way about those stranded. Moreover, a surprising 65 percent of whites blamed city residents themselves and the mayor for those who were trapped in New Orleans, whereas just 20 percent blamed the Bush administration.[20]

Such responses could be viewed as proof of Adolph Reed's argument that the neoliberal mantra of limited government, far more than any forces of racial discrimination, should be privileged in any analysis of the ideological conditions underlying the Katrina tragedy.[21] There is reason to believe, however, that these arguments cannot be applied in the same manner to both blacks and whites. Whereas African Americans surveyed about such questions tended to view their meaning in broad, systemic terms, as referenda on the existence of racial inequalities in government institutions and policies, whites tended to interpret them rather narrowly and, in some sense, superficially— as asking whether they believed that overtly racist attitudes by government officials actually caused them to intentionally allow blacks to suffer.[22] In a sense, whites, much more than blacks, seemed to exhibit a kind of neoliberal reflex to shrink the broader problem of racial inequality down to the level of individual actions—a tendency that reflected the initial reactions of Bush administration officials. For example, in the face of the rising crescendo of criticism from black leaders on September 2, Secretary of State Condoleezza Rice exclaimed, "That Americans would somehow in a color-affected way decide who to help and who not to help, I, I just don't believe it." Then, several days later, President Bush echoed this same sentiment when he claimed, "When those Coast Guard choppers . . . were pulling people off roofs, they didn't check the color of a person's skin."[23]

These expressions of faith in color-blindness amount to a tendency to minimize racism that Eduardo Bonilla-Silva has identified as a key component of an ideological form he refers to as "color-blind racism."[24] Born out of the struggle between race-consciousness and color-blindness during the era of minority empowerment and white backlash in the 1970s and 1980s, this reflex to minimize the existence of racism is based on the conviction that racial parity has been achieved. While few observers viewed it this way at the time, this move to embrace color-blindness and deny racism was inscribed within the broader neoliberal turn that was underway during the Reagan years. The rise of neoliberalism led to, as David Theo Goldberg has argued, "an increasing stress on individualized merit and ability in the name of racelessness."[25] And yet, in spite of all this, cultural racism flourished during these years, especially within the context of the Republican Party's anti-welfare campaign, which was at the symbolic core of the Reagan Revolution, and it flourished because its services were needed to confront the realities of the American racial order. The spectacle of black ghetto poverty, which, regardless of the number of poor whites who found themselves stranded alongside blacks, was once again on vivid display after Hurricane Katrina, poses some serious legitimacy problems for color-blindness. Thus, when faced with hard evidence of racial inequality, color-blindness gives way to cultural racism—in the form of discourses focused on individual and group shortcomings—to explain the circumstances at hand. In short, when the myth of color-blindness is faced with undeniable racial inequalities, the move to cultural racism works to deflect attention away from any consideration of the structural and historical conditions behind these injustices. In this sense, it possesses the same individualizing logic as neoliberalism.

That this process was at work in white civil society in the aftermath of Katrina can be glimpsed in the survey data indicating how whites felt about those left behind—particularly the alarmingly high percentage who blamed New Orleans residents themselves for not evacuating and who expressed limited concern for the flood victims.[26] No less illuminating is the finding by one survey that one out of two whites believed that those who "entered stores and took things in the first few days after the hurricane" were looting rather than searching for vital supplies.[27] This perception of blacks as looters, it must be pointed out, owed much to the almost pornographic imagery of such activities broadcast by the mainstream media, as well as the widely circulated expressions of righteous anger voiced by Louisiana Governor Kathleen Blanco and a range of television and radio personalities. Emblematic of such coverage is CNN host Nancy Grace's description of the scene in New Orleans: "Armed looters roam the streets of New Orleans, ransacking stores in broad daylight, stealing everything in sight—electronics, food, clothing and guns—carjackers attacking vehicles leaving the city, multiple shootings. . . ."[28] And yet, this was not a case of a perspective imposed on an unwilling public; in the production of the looting frame, which was enhanced by unsubstantiated rumors of rape and indiscriminate murder, the media was reflecting back on American society the figments of cultural racism that had inhabited the white mind for decades.

At the core of such views are notions that crystallized and hardened during the decades-long campaign for welfare reform that began during the Great Society backlash of the 1960s, when the idea of a welfare-dependent and undeserving poor black population was paraded across the public stage in a number of different costumes—welfare queens, poverty pimps, sexual predators, and gangbangers. The politics of welfare reform during the 1970s and 1980s became central to the Republican Party and the conservative ascendency, providing a flow of "hot-button" issues and images that provided the emotional and ideological underpinnings for grassroots anti-tax and anti-integration movements. If in 1968, the Kerner Commission declared white society to be deeply implicated in the development of "two societies, one black, one white," these movements represented the progressive disengagement of whites from the problem of racial inequality. A range of neoliberal discourses—individualism, meritocracy, property and consumer rights—took shape to justify this move by whites to effectively privatize their lives, but such a limited view of civic duty required the supplement of cultural racism; it was also necessary to believe that those who needed help were incapable of benefiting from it. Thus, neoliberalism and cultural racism worked in tandem throughout the post–civil rights era, and they were doing so as whites gazed at flood-ravaged New Orleans and asked "why didn't they leave" rather than "why didn't the government make sure to evacuate them." What was transpiring from an ideological standpoint resembled what Mahmood Mamdani has referred to as the "culturization of politics"—the transfer of political acts and events onto the terrain of culture, where they become dissociated from questions of structure, power, and, ultimately, political mobilization.[29] As stranded New Orleans residents—many of them elderly and without the resources to escape—transformed into looters, gangbangers, and welfare dependents, culture, not structure, filled the frame, and blame shifted over to the victim yet again.

Black Community and Leadership in a Neoliberal Age

Scholars and journalists alike have made much of the dramatic differences between black and white survey responses to Hurricane Katrina. Undeniably, there is some strong evidence among this data to support the conclusion that a broad consensus characterized the black understanding of the event as one of great racial injustice. An impressive 84 percent of African American respondents in the Pew survey, for example, agreed with the statement that the government response would have been faster had the victims been white.[30] And yet, it is possible to overstate the case for black consensus, and by doing so to overlook some possible indications of critical fractures in the black "community"—in particular, class divisions that may tell us something more about why black outrage did not develop into protest politics.

A *Newsweek* magazine poll conducted by Princeton Survey Research Associates International appearing on September 29, 2005, found that 65 percent of "non-whites" believed that government was slow to rescue those trapped in New Orleans because they were black. While this figure appears high in comparison to the 36 percent of whites who gave the same response, one could argue that is not nearly high enough to indicate anything resembling a clear black consensus. Indeed, questioned in the emotional heat of the moment, when the media was promoting a racial injustice frame, a full 35 percent of blacks nonetheless claimed not to view race as playing a role in the rescue debacle.[31] While a class analysis of this figure must remain a matter of speculation since participants in this poll did not provide socioeconomic or occupational data, it is likely that the percentage of middle-class black respondents rejecting a racial interpretation was significantly higher—perhaps even more than 50 percent. This is not the only evidence of likely divisions in the black community. Another survey, conducted by researchers in October and November of 2005, revealed that nearly one in three blacks did not feel "very sympathetic" with those left behind, that more than one in three did not think that flood victims "deserve[d] . . . a great deal of assistance," and that nearly as many blacks blamed the Bush administration (40 percent) as they did "the mayor and local residents" (38 percent) for those "trapped" in the city.[32] Moreover, a CBS News poll conducted six months after Katrina found that more than 40 percent of African Americans thought that race was either "not a factor" (24 percent) or a "minor factor" (17 percent) in the "rebuilding progress" of the city.[33] These numbers suggest a reluctance on the part of a significant (one in three, on average) segment of the black community to embrace a clear-cut racial interpretation of the events following Hurricane Katrina. While interpreting these figures in this way might appear to be a question of seeing the glass half-empty rather than half-full, the context of black political quiescence in the aftermath of Katrina suggests we take them seriously as indicators of incertitude about the racial meaning of Katrina within some sectors of the black community.

Beyond the mere fact that the social standing of the middle class—whether black or white—makes it more prone to invest in notions of color-blind meritocracy and self-help, the discursive struggles raging among African Americans over the state of the black poor from the 1990s onward suggest that a significant portion of the black middle class was less susceptible to interpreting Katrina within a racial injustice frame. Once again, this does not at all imply that black middle-class professionals were not active in the aftermath of the hurricane. Black churches raised millions of dollars; the NUL cosponsored a Relief Telethon on Black Entertainment Television; black musicians, including popular hip-hop artists and jazz greats alike, rallied to raise funds for flood victims; and thousands of black college students and activists descended on New Orleans to do their part. And yet, as Michael Eric Dyson asserts, "Charity is no substitute for justice."[34] As was the case in white civil society, political engagement for African Americans immediately took the form of relief and charitable activities, a situation that worked to soften emotions of injustice and deflect attention away from questions of accountability. Blacks were, like whites, constrained by a neoliberal framework that pointed them towards individual and private solutions to what was a public problem.

This was not the only manner in which neoliberal rationalities constrained contentious political action. As Michael Dawson has observed, "Black leadership has bought into the neoliberal ideology that defines organizational activity as lobbying, which is built on individualist leadership models and emphasizes civil society as the sole route to group advancement."[35] This lobbying approach was at work when the chairman of the Congressional Black Caucus, Melvin Watt, used his appearance on CNN to unequivocally refute black political leaders who were alleging racism as New Orleans residents were still stranded on rooftops. In a sound bite that could have easily come from the mouth of President Bush, Watt asserted, "I don't think God visited this hurricane only on black people; he visited it on New Orleans, and the people who were in New Orleans—black, white, or otherwise—were victims of it."[36] And the lobbying model of political action was no less in evidence when Bruce S. Gordon, president of the NAACP and a former telecommunications executive, met with President Bush in the Oval Office to convince him to come address the NAACP for the first time in June of 2006. Prior to making overtures to President Bush, Gordon, it should also be mentioned, used his ties to corporate America to effectively lobby for relief assistance.

To be sure, the NAACP played a key role in the relief effort and in defending the rights of flood victims, and its chairman, Julian Bond, did, in his address to the ninety-eighth annual NAACP convention in 2007, compare the reconstruction of New Orleans to a "lynching." But, in the end, lobbying and relief defined the NAACP's role, and as such, it ended up contributing to the depoliticization of Hurricane Katrina in black civil society. Indicative of the orientation of the NAACP away from a posture that could promote the kind of political mobilization capable of challenging the racial status quo was the address of NAACP president Benjamin Todd Jealous on Hurricane Katrina's five-year anniversary. Entitled "Katrina's Lessons Still Not Learned Five Years Later," the "lesson" that Jealous highlighted, above all, was that the "US overdependence on fossil fuels is wreaking deadly short and long term destruction." While Jealous does indeed point to the ways in which such trends disproportionately affect African American coastal communities, this move towards a greening of black progressive politics is once again more geared towards lobbying than grassroots mobilization.[37]

The NAACP's stance on Katrina must be understood not only in the way it reflects a lobbying model of political action, but also in how it relates to its efforts at maintaining its own predominantly middle-class constituency. The class location of this constituency, which gives it much greater exposure to white civil society, makes it more invested in a color-blind vision of the world, and less comfortable with a political position that can be construed as identity-driven. The charge of "playing the race card" has become a potent discursive weapon for the American right in the Katrina era, and, as the Obama campaign revealed, this label must be avoided at all costs if one does not wish to be marginalized from mainstream politics as a racial *provocateur*. Hence, even as some black leaders were, in the immediate aftermath of the storm, voicing concerns about race, others like Watt were either refusing the racial perspective or choosing their words carefully. Even Jesse Jackson, whose reputation as someone who "plays the race card" is irreversible, hedged on the issue at first, claiming that race was "at least a factor." Therefore, in a sense, the black reaction to Katrina seemed to reveal that the United States, in the bipartisan "us versus them" color-blind era of Katrina, has become more like republican France in its increasingly hostile predisposition towards racial identity-based politics. Yet, whereas in France this republican frame at least holds the potential of facilitating a "social" or class frame that can serve a key role in redistributive politics, such universalism in the guise of color-blindness within the neoliberal political culture of the United States is a lose-lose situation. And it is here, it cannot be emphasized enough, that white civil society could have played a decisive role. The expression of even symbolic support for the cause of racial justice would have provided critical cover for black organizations.

Hence, if Katrina were going to spark contentious political activities within black civil society, middle-class African Americans would have had to have "played the race card." But the "race card," it seems, has become a bit too "ghetto." In other words, for many members of the black middle class, the very notion of identity-based mobilization now carries with it the class and cultural resonances of the era of ghetto rebellions and struggles around welfare entitlements. In fact, one could read the Million Man March that took place a little more than a decade before Katrina as a middle-class attempt to dissociate black politics from such ghetto residues and resituate black political engagement within a middle-class context. That the two most important values that defined the march were "self-sufficiency" and "personal responsibility" suggests that, much like for its white counterpart, the ideological terrain of black civil society has been profoundly shaped by the interplay of neoliberalism and cultural racism. The embrace of such values by the conservative-leaning segment of the black middle class implied—consciously or subconsciously—the tacit acceptance of the same ghetto demons that had white Americans rallying around welfare reform.

While the Million Man March may have seemed like a distant memory by the time Katrina made landfall, its middle-class critique of the pathologies of black ghetto culture had outlived the culture wars of the 1990s.[38] In 2004, for example, television personality Bill Cosby sparked a widely publicized debate within the black community when he blamed high black dropout rates on dysfunctional parenting. While some black leaders and opinion-makers criticized Cosby for blaming the victim and underplaying the forces of institutional racism, others praised his candor. One of those defending Cosby, for example, was black Pulitzer Prize–winning columnist Cynthia Tucker, who in May of 2004 claimed, "at the dawn of the 21st century, personal responsibility has at least as much to do with success in modern America as race, especially since the Supreme Court decision in *Brown v. Board* rolled back much of systemic racism."[39] Widely recognized as a liberal, Tucker's support of such views indicates that blacks rallying around notions of "personal responsibility" hardly represented a hardcore conservative fringe. In fact, towards the beginning of his presidential campaign, Barack Obama chose this same register as he sought to situate himself in the political center of the black community. In his famous speech in Selma, Alabama, for example, Obama told a mostly black audience that the black struggle was "90 percent of the way there," criticized "daddies [for] not acting like daddies," and poked fun at an imaginary "cousin Pookie" for not getting off the couch.

That Obama could make such a speech merely a year and a half after Katrina hit is evidence of how much the notions of dysfunctional ghetto culture had weathered the storm.[40] Taken together with survey data suggesting that nearly as many blacks blamed the federal government as the "mayor and local residents" for those trapped in New Orleans, it is reasonable to conclude that substantial numbers of middle-class blacks were also asking "why didn't they leave" rather than "why didn't the government evacuate them." Indeed, the "culturization of politics" was at work in black civil society as well, a situation that kept black mobilization within the domain of relief rather than political contestation. Hence, an event that could have challenged such cultural perspectives on structural poverty and systemic racism may have ended up, in the end, reinforcing them. As the city of New Orleans debated a reconstruction program that sought to eliminate public housing and hand the city's future over to developers, black City Council president Oliver Thomas defended such policies by publicly stating that past government programs have "pampered" poor people, and telling public housing residents that they should not return unless they are willing to work. Capturing decades of neoracist blame-the-victim, anti-welfare sentiments, Thomas declared: "We don't need soap opera watchers right now." This was the culturization of politics in the service of neoliberalism, *par excellence.* And, while such comments were widely reported in the mainstream press, they barely caused a stir within black civil society or the black counterpublic.

Thus, aside from all of the other things that Hurricane Katrina seemed to suddenly reveal about race and politics in the United States, it demonstrated how normalized ideas of personal responsibility and the black poor's lack thereof have become within the black community as well. One of the reasons for this is that the black middle class, not unlike its white analogue, has, since the 1980s, required the services of rationalities that can do the daily psychological labor of justifying the increasing gap between itself and the black poor. In the case of the black middle class, moreover, sensibilities of racial solidarity further complicate such rationalizations. As Mary Patillo has shown in her brilliant study of black class relations in a gentrifying neighborhood in Chicago, the racial ties that bind professional and working-class African Americans in such neighborhoods are put to the test in community struggles over charter schools, public housing, and the "proper" use of public space. However, if, as Patillo argues, middle-class blacks in these situations envision themselves as occupying leading roles in a project of racial uplift within the larger black community, they also cannot help observing that, in class-defined struggles over limited neighborhood resources in the neoliberal city, they are almost always the winners and their poorer neighbors the losers, and this understanding, in turn, seeks comfort in the perceived cultural short-comings of the lower class.[41] And here we see again—this time from a bottom-up perspective—how much neoliberalism depends upon the culturization of politics it plays a leading role in bringing about. What Patillo's study shows is that the neoliberal turn has met some resistance in the black community, but, as Hurricane Katrina so poignantly revealed, it is well on the way to affecting the same depoliticizing processes it has already accomplished in white civil society. That this culturization of politics is, in the case of the black community, moved forward by elected black officials like Oliver Thomas and Ray Nagin, and in the case of the white community, pushed beneath the surface of public discourse by the seemingly ubiquitous power of color-blindness makes it a hard nut to crack for those still interested in redistributive political projects.

Notes

1. CNN Transcript, *American Morning,* September 2, 2005, 7:00; CNN Transcript, *Lou Dobbs Tonight,* September 2, 2005, 18:00.

2. William A. Gamson, *Talking Politics* (Cambridge, U.K.: Cambridge University Press, 1992), 32.

3. David A. Snow, "Framing Processes, Ideology, and Discursive Fields," in *The Blackwell Companion to Social Movements,* ed. David A. Snow et al. (Malden, Mass.: Blackwell Publishing Ltd., 2004), 380–412.

4. CNN Transcript, *Lou Dobbs Tonight,* September 2, 2005, 18:00.

5. Robert C. Lieberman, "'The Storm Didn't Discriminate': Katrina and the Politics of Color Blindness," *Du Bois Review: Social Science Research on Race* 3, no. 1 (2006): 7.

6. See, for example, Matthew D. Lassiter, *The Silent Majority: Suburban Politics in the Sunbelt South* (Princeton, N.J.: Princeton University Press, 2007).

7. George W. Bush, "Address to the Nation on Hurricane Katrina Recovery from New Orleans," Louisiana, September 15, 2005, Weekly Compilation of Presidential Documents, Administration of George W. Bush, 2005, 1405–9.

8. Adolph Reed, "Class-ifying the Hurricane," *The Nation,* October 3, 2005: www.thenation.com/article/class-ifying-hurricane (accessed March 4, 2014).

9. This incident was covered on December 18, 2005, by the widely viewed CBS news show *60 Minutes.*

10. Michael B. Katz, "Why Don't American Cities Burn Very Often," *Journal of Urban History* 34 (2008): 186.

11. Pew Research Center, "Two-in-Three Critical of Bush's Relief Efforts: Huge Racial Divide over Katrina and its Consequences," Pew Research Center for the People and the Press, 2005, www.people-press.org/2005/09/08/two-in-three-critical-of-bushs-relief-efforts/ (accessed March 4, 2014).

12. Robert D. Putnam, *Bowling Alone: The Collapse and Revival of American Community* (New York: Simon & Schuster, 2001).

13. While the Million Man March's emphasis on "personal responsibility" and "self-help" makes it a somewhat complicated example of mass-based political protest, one must not forget that it responded to the Republican Party's victory in the 1994 Congressional election, and, more specifically, to its "Contract with America."

14. I borrow the term "rationality" from Wendy Brown, who defines it as "a specific form of normative political reason organizing the political sphere, governance practices, and citizenship"; see Wendy Brown, "American Nightmare: Neoliberalism, Neoconservatism, and De-Democratization," *Political Theory* 34 (2006): 693.

15. My notion of a "counterpublic" is based on Nancy Fraser's description of "subaltern counterpublics" as "parallel discursive arenas where members of subordinated social groups invent and circulate counterdiscourses to formulate oppositional interpretations of their identities, interests, and needs." See Nancy Fraser, "Rethinking the Public Sphere: A Contribution to the Critique of Actually Existing Democracy," *Social Text* 25–26 (1990): 56–80.

16. Michael C. Dawson, "After the Deluge: Publics and Publicity in Katrina's Wake," *Du Bois Review* 3, no. 1 (2006): 240.

17. Quoted in Dawson, "After the Deluge," 247.

18. The Common Ground Collective, founded by former Black Panther Malik Rahim, mobilized more than ten thousand activists and college students. See Rachel E. Luft, "Beyond Disaster Exceptionalism: Social Movement Developments in New Orleans after Hurricane Katrina," *American Quarterly* 61 (September 2009): 502.

19. Brown, "American Nightmare," 692.

20. Leonie Huddy and Stanley Feldman, "Worlds Apart: Blacks and Whites React to Hurricane Katrina," *Du Bois Review* 3, no. 1 (2006): 105, 108.

21. Reed's writings on the subject, which initially appeared in *The Nation* and *In These Times* in 2006, are compiled in Adolph Reed Jr., "Class Inequality, Liberal Bad Faith, and Neoliberalism: The True Disaster of Katrina," in *Capitalizing on Catastrophe: Neoliberal Strategies on Disaster Reconstruction,* ed. Nandini Gunewardena et al. (Plymouth, U.K.: Altamira Press, 2008), 147–54.

22. On the differences in black and white views of these questions, see Paul Frymer et al., "New Orleans Is Not the Exception: Re-Politicizing the Study of Racial Inequality," *Du Bois Review* 3, no. 1 (2006): 42.

23. Both Bush and Rice were, in part, responding to allegations made by the NAACP's Washington Bureau director, Hilary Shelton, that rescue workers had taken away "boatloads" of white residents while blacks were left stranded.

24. Eduardo Bonilla-Silva, *Racism without Racists: Color-Blind Racism and the Persistence of Racial Inequality in America* (New York: Rowman & Littlefield, 2009).

25. David Theo Goldberg, *The Threat of Race: Reflections on Racial Neoliberalism* (Malden, Mass.: Wiley-Blackwell, 2009), 331.

26. In addition to the survey of Huddy and Feldman, a CBS News poll conducted shortly after

the hurricane showed that people blamed the "situation in New Orleans" as much on the residents themselves as all levels of government.

27. Huddy and Feldman, "Worlds Apart," 104.

28. CNN Transcript, *Nancy Grace,* August 31, 2005, 20:00.

29. See Mahmood Mamdani, *Good Muslim, Bad Muslim: America, the Cold War, and the Roots of Terror* (New York: Pantheon, 2004).

30. Pew Research Center, "Two-in-Three Critical of Bush's Relief Efforts."

31. An NBC/Wall Street Journal poll revealed that three in ten blacks disagreed with the statement that the Bush administration would have responded faster had the victims been in white suburbs.

32. Huddy and Feldman, "Worlds Apart," 103–8.

33. CBS News Poll, Monday, February 27, 2006.

34. Michael Eric Dyson, *Come Hell or High Water: Hurricane Katrina and the Color of Disaster* (New York: Basic Civitas, 2005), 152.

35. Dawson, "After the Deluge," 247.

36. CNN Transcript, *Lou Dobbs Tonight,* September 2, 2005, 18:00.

37. See www.naacp.org/news/entry/five-years-after-katrina-lessons-still-not-learned/.

38. For a discussion of the 1990s "culture wars," see Robin D. G. Kelley, *Yo' Mama's Disfunktional: Fighting the Culture Wars in Urban America* (Boston: Beacon Press, 1997).

39. Cited in Dyson, *Come Hell or High Water,* 148.

40. In this speech Obama did discuss New Orleans, arguing that residents there suffered from a "hope gap" before the storm even hit.

41. Mary Patillo, *Black on the Block: The Politics of Race and Class in the City* (Chicago: University of Chicago Press, 2007).

Reformers, Preservationists, Patients, and Planners

Embodied Histories and Charitable Populism in the Post-Disaster Controversy over a Public Hospital

ANNE M. LOVELL

On the morning of September 28, 2005, when Viola Green lowered her legs from her bed and wiggled her toes to find her slippers, she felt instead something warm and wet. A pool of water ankle-deep covered the bedroom floor.[1] She waded through to the living room to open the front door. What she saw made her quickly check that her ladder was still leaning where it belonged. As she later explained to me, "I had a brother—he dead now—he always say, 'take a ladder and put it at the front of the door. Inside a wall where you can get it, and it won't ever fall down.'" She hollered out the door for help: "Mr. Bailey across the street, he called out: 'Miz Viola, you alright?' 'No, I'm not alright.' My voice echoed [like] across a big lake. I told him to get the lady live in the third house. The lady with a cell phone."

Viola Green then hoisted her heavy, seventy-six-year-old diabetic body to the top of the ladder. She was nervous, afraid of falling if she looked down. By the time Frances, the lady with the cell phone, arrived hours later, the water was almost up to Viola's throat. Some men, neighbors, eased her down, out the door, and into a boat. They rowed her to a nearby new but unoccupied two-story house. Other neighbors had broken down the door and were helping elderly and sick to the top floor. Mr. Jackson was among them. He was the old man whose house Frances cleaned and whose medical forms she filled out, as he couldn't read or write. This was supposed to be his dialysis day, and he was in a bad way. Frances meanwhile sent some teenagers for water and food from the boarded-up corner store. "We had a senior citizen center on that second floor," Frances recalled later. It would be hours before neighbors completed their rescues and finally started moving the sickest to a hospital one parish over, and the others to the causeway.[2]

This brief Katrina narrative about a mostly African American neighborhood highlights two issues about disasters. On a general level, it illustrates the resiliency, inventiveness, and mutual aid in their aftermath, those "paradises built in hell" so compellingly conveyed by Rebecca Solnit. "Acts of collective creation," Émile Durkheim called them, moments during upheavals when normal forms of governance disappear and people "are brought into more intimate relations with one another, when meetings and assemblies are more frequent, relationships more solid and the exchange of ideas more active." The image of neighbors tending to individual bodies—sick, disabled, or elderly—evokes a social body, not as a symbolic dimension of single bodies, but as the concerted action of many individuals, aware at least momentarily of one another's capacities, limitations, and common fate.[3]

More specifically, the narrative suggests the omnipresence of illness and disability in poor neighborhoods *before* Katrina, a point often lost in humanitarian appeals and master narratives of disaster-induced post-traumatic stress conditions. Then, as now, everyday sociability in places like Viola's neighborhood often takes place on porches and stoops. People will call out to one another from wheelchairs, leaning on canes or while breathing from an oxygen hook-up, and talk sometimes drifts to the "shut-ins" down the street or to someone's recent passing away. This interweaving of disability, disease, and poverty, so manifest even among the young in public and quasi-public space, translates into dramatic statistics that place New Orleans and Louisiana near or at the bottom of all the states on most health indicators, including for chronic diseases of poverty-advanced diabetes, obesity, high blood pressure, cardiovascular disease, HIV, and AIDS. "Disease's double" is how a pioneering Italian social psychiatrist identified the social inequality behind illness and disability. Applied to Katrina, this idea extends the analysis of preexisting power relations revealed by disasters to include the interweaving of race and class with disease and disability, the body politic and individual bodies.[4]

A multivoiced social movement arose around one of the major controversies of the recovery process—the closure and proposed replacement of the city's major public medical facility, Charity Hospital—that came to symbolize the health-care void citizens like Viola and her neighbors faced after Katrina. The hurricane and surge from the socially engineered disaster that flooded 80 percent of the city also destroyed most of the clinics and hospitals in Greater New Orleans and spurred the evacuation of healthcare workers, including the largest single displacement of medical doctors in U.S. history.[5] In Katrina's aftermath, the health-care void offered a ripe terrain for modernizing and rationalizing health care, bringing to mind the so-called health-policy laboratories that had found bipartisan support in several states after the failure of President Clinton's national health reform efforts.[6] In New Orleans, such experimentality materialized through both grassroots and formal, top-down efforts, producing two narratives about health deprivation—one centered on Charity Hospital, and the other on medical homes and primary care. While both narratives are shaped by health-care reform efforts underway in the years preceding Katrina, Charity's history is embedded in colonial New Orleans and in twentieth-century Louisiana populism.

Historical racialization and charitable populism are embodied in a form of situational identity with Charity Hospital, expressed through the moniker "Charity Hospital Babies." Pre-Katrina health care had its cycles of reform, one hospital-centered and the other promoting a vision of community-based facilities; post-storm recovery efforts inherited both. Charity Hospital along with proposed alternatives transformed the idea of the hospital into a boundary object for communities of practice beyond the health arena. Charity's demise and connected, contested events are framed by competing narratives anchored in the social movement to save Charity and in the primary-care experimentality that the disaster spurred.[7]

The Racial Construction of Charity Hospital, Charitable Populism, and Situational Identity

In a terrifying drama, the storm and ensuing disaster flooded the basement and damaged other floors at New Orleans's Charity Hospital, one of the largest and oldest public hospitals in the United States. One week after medical staff and surviving patients were finally evacuated, Lieutenant General Russel Honoré, a late appointee for overseeing military relief efforts in the chaotic city, sent two hundred specialists from the military relief workforce, joined by volunteer doctors and nurses, to drain, clean, decontaminate, and repair enough of the building to reopen three floors. Shortly afterwards, Charity's administrators urged Governor Kathleen Blanco to order the facility closed, without seeking legally required legislative action. In the aftermath, the Louisiana State University Healthcare Services Division (LSUHSD) laid off almost three thousand employees, many of them African American. The closure disrupted medical education, severely cut health-care expenditures for the uninsured, and left patients to fend for themselves. Honoré, perhaps the only popular official figure in Katrina's disaster response, expressed shock at the decision.[8] "I still get mad every time I look at Charity Hospital," he would later reflect. "In the culture of preparedness, if the first floor of that building had been a parking garage, the hospital would still be open."[9] Medical staff meanwhile opened makeshift services in tents and closed-down buildings, dubbing themselves "The Spirit of Charity." Months later, a smaller version of Charity's sister hospital, University, opened. Charity loomed empty against the urban skyline.[10]

Closing Charity as irreparably damaged—a claim challenged by some who witnessed the repairs[11]—provided LSUHSD grounds for requesting FEMA funds for a new hospital that, I would later learn, was conceived before Katrina. In March 2006, doctors and nurses rallied to reopen Charity, joined by advocates from the pre-Katrina Louisiana Health Care Community Coalition, the Southern Christian Leadership Conference, ACORN, the New Orleans teachers' union, progressive Catholics, People's Hurricane Relief Fund, and other organizations. The Committee to Reopen Charity sprouted from that event. Besides calling to reopen Charity, advocates fought potential damage from a new medical complex in the Lower Mid-City neighborhood where Louisiana State University and the U.S. Department of Veteran Affairs (VA) announced they were building replacement facilities, with joint maintenance, clinical, and other departments. The hospital controversy played out over the next few years in class-action lawsuits; city council and state legislative hearings;[12] public forums; the media; environmental impact, site selection, and building review processes; and neighborhood-recovery planning meetings. Meanwhile, former Charity patients and new ones filled private hospital emergency rooms in adjoining parishes, traveled across the state and region for care, or went without. The city's Health Department reported higher-than-expected mortality rates in the years following Katrina. Yet the tenacity of the Charity Hospital controversy throughout the recovery years, when residents were struggling with homes, jobs, broken networks, and health, begins with a historical legacy.[13]

Slavery to Civil Rights at Charity Hospital

The relationship of New Orleanians to Charity Hospital is culturally embedded in a racialized and sometimes racist reading of its history but also a deeply rooted Louisiana populism. Generations were born, treated, and died at Charity Hospital, where many African American parents, siblings, and extended kin found lifetime employment. Charity Hospital generates a powerful source of collective identity I call "situational" (in reference to *site* but also *situations,* or particular moments in the life-course), crucial to the post-Katrina movement for reopening.[14]

Antebellum Charity Hospital depended on the labor of enslaved African Americans, including highly skilled surgeons and phlebotomists, sometimes claimed as the first African American physicians. Until the Civil War, Charity served European-descended patients. After Emancipation and the disappearance of plantation-based sick-houses, and with urban charitable institutions diminishing, black patients turned to Charity Hospital. By the 1900s, most Charity patients were African American. But the early twentieth-century presence of African American physicians at Charity Hospital has been obfuscated in written histories in which physicians' "whiteness" is the unmarked category. Nationally, the percentage of African American physicians steadily declined, never recouping its all-time high of 1910.[15]

Throughout the first half of the twentieth century, Charity Hospital gained national recognition for first-class care and medical research. But its accomplishments depended on African American bodies for what by today's standards were ethically tainted experiments. For example, in 1940, Tulane Medical School studied how quickly the body processes radioactivity on three hundred mostly black female Charity patients, uninformed that they were undergoing risky procedures. The women were injected with or ingested radiation equivalent to up to a hundred chest x-rays. Additionally, many new drugs were tested on Charity patients, including for rare diseases like leptospirosis ("canefield fever"), tularemia, and other rodent-borne ailments unknown elsewhere in the United States. Medical researchers availed themselves of unclaimed cadavers, and rumor held that African American patients at Charity were commonly given cascara and magnesia in a notorious black bottle to hasten their death and provide further corpses.[16]

Until 1965, African Americans could receive medical care in New Orleans only at the historically black private hospital, Flint-Goodridge, or in Charity Hospital's segregated facilities. After the Civil Rights Act, the hospital desegregated patient care, but integration of personnel languished. Charity thus provided a major focal point in African American struggles, from civil rights to the War on Poverty and black power. Civil rights leaders fought well into the 1970s to integrate Charity staff, ensure representation at upper levels, demand responsiveness to African American communities, and prevent budget cuts and clinic closings. Celebrated figures like Oretha Castle Haley (1939–

1987), deputy director to the hospital's first African American administrator, and the Reverend Avery A. Alexander (1910–1999), after whom the hospital was renamed in 1991, led many of these struggles.[17]

Populism or Charity?

If racial readings[18] of Charity Hospital persist in the city's social imagination, populist understandings also influence situational identity with Charity. Huey P. Long's governorship (1928–32) and Share Our Wealth program promulgated the principle that all Louisianans should benefit from health care and that even "poor boys with good records" could become doctors.[19] The Long dynasty's power was such that even political opponents supported state-funded social services. Long established the LSU medical school at Charity in 1930. After Long's assassination, Louisiana implemented his plan for building a system of charity hospitals throughout the state, providing free health care to anyone who could not pay. In New Orleans, a new Art Deco building replaced the older Charity Hospital in 1938. Well into the 1960s, when most other states were dismantling public hospitals, Louisiana continued building charity hospitals, an exception that attests to the endurance of its populist policy.

Yet Louisianans perceive medical care for those unable to pay as charity, rather than as tax-generated. This confusion is not surprising. Historically, Charity Hospital doubled as an almshouse, with inpatients staying up to a year as late as the 1930s, from economic, not medical need. Charity also sheltered homeless throughout the Great Depression. The 1938 Works Progress Administration guide to New Orleans describes the hospital as state-operated for the benefit of "indigent citizens of Louisiana" and having provided free hospitalization to "poor people of New Orleans" since its founding in 1736. It notes that all New Orleans doctors and many practicing in other parishes donate a percentage of their time weekly to the hospital and its free clinics, "for both white and colored." The Sisters of Charity ran the nursing and building services with free labor in accordance with their vows until the late 1940s; sister-nurses remained until the late 1980s.[20]

As the hospital became increasingly dependent on Medicaid in the 1960s, deeply engrained notions of "Charity" as a gift gave way to racist connotations of charity as welfare and Charity Hospital as a delivery room for "welfare babies." To counter stigma, the Louisiana Healthcare Authority (LHA), which took over administration of charity hospitals in 1990, recommended dropping "Charity" from its official title. LHA also emphasized that charity hospitals provided medical education for physicians who served all Louisianans, poor or not. But the "Charity" name and stereotypes persisted.[21]

For many African Americans, working poor, and even some middle-class New Orleanians, Charity Hospital provided certainty of medical care, regardless of long waits.[22] Although U.S. emergency rooms provide safety nets, Louisiana's particularity was the fusion of the politics of race and free care in a *charitable populism* that underlies situational identity with Charity Hospital.

Charity Hospital Babies as Situational Identity

Situational identity stems from a sense of place attachment that fuses concreteness of place with subjective desires and intentions. Artifacts and place-based stories contribute to the narrative construction of identity, itself impossible without spatial and temporal referents. Not quite a home (in the ontological sense), Charity Hospital provided a *temporary* place or way station, symbolically akin to cemeteries, schools, and other sites that generate identity by marking stages in the life cycle. But, like many homes, Charity was imbued with agency, familiar and gendered. Physicians who trained there often referred to it as "'Mother Charity." She was, as one physician-historian wrote, "'the great stone womb' that gives birth to multiple offspring and becomes 'the great stone breast' that nurtures them all their lives."[23]

Nowhere is the maternal metaphor stronger than in the moniker "Charity Hospital Babies," made famous by the late rhythm-and-blues singer Ernie K-Doe and adopted as the rallying cry by advocates for re-opening Charity Hospital after Katrina. According to an ex–Charity Hospital director and a sociologist, the nickname stands for the hospital's legacy of providing affordable healthcare for all: "As a public service, thousands of babies were delivered in the charity hospitals and many residents today proudly proclaim that they were a 'charity' baby." "Charity Hospital Babies" comprises a discursive unit legitimizing those who embrace and understand it, but actively excluding outsiders to whom the expression remains opaque. It indicates a certain kind of New Orleanian, much as does the street or neighborhood where someone lives,[24] thus conferring intense familiarity, group belonging, and distinction within the larger social space of New Orleans.[25]

Charity Hospital, then, embodies the familiar quality attributed to both mothers and home—the opposite of Freud's uncanny (*umheimlich*, literally un-homelike)—translated into a sense of stability and familiarity for patients whose lives were rendered chaotic from decades of gentrification and displacement, neoliberal economic policies, and racism. "Code Blue," a thirteen-part documentary about Charity's emergency services, repeatedly illustrates its homelike, maternal quality, as doctors and nurses attend church with a former patient, lecture a gunshot victim on violence, or guide schoolchildren through the services "where you will be coming back your whole life," reinforcing, if not forcing, the sense of belonging. One former Charity patient and now investigative journalist condensed this maternal metaphor in a public policy forum: "[T]hree days before Katrina, I wanted to retrieve my records from Charity's record room. . . . All these mothers were coming in to get their kids' medical records to enroll them in school. And it really was a profound moment when I realized that this is the one piece of stability in very chaotic lives. [Charity] really functioned as the womb of the city, and I don't think you can begin to grasp the loss until you recognize that it was the mother of the city, and our mother was shut down. . . . [If] you talk about replicating my mother, or the place that offered me solace, hope and life, make sure you've got another mother in place. Don't talk about having another mother in place in five years, because I'll be dead by then."[26]

This sometimes intangible cultural identity was not lost on proponents of closing Charity post-Katrina. In 2006, the LSU Health Sciences Center chancellor acknowledged the attachment to Charity in debates over a new hospital, noting that "many people have difficulty letting go of what was there" but dismissing the vision behind it as "a New Orleans that was not there." The city-appointed Bring New Orleans Back (BNOB) commission cautioned through its Health Committee recommendations against a "one-size-fits-all" approach to reforming health care, given the city's racial and cultural diversity. Even the Public Affairs Research council (PAR), a critic of state health policy, admitted that "public understanding and support" would be needed to replace the Charity system.[27]

Before the Storm and Afterwards: Two-Tiered Care and Cycles of Reform

During the decade before Katrina, in a national climate of health rationalization, growing evidence-based medicine, and cost containment, Louisiana health reformers rethought the human and economic costs of what was in effect a two-tiered system of health-care provision. In New Orleans, the two LSU-run teaching hospitals, Charity and University, comprised the lower tier, and fourteen private for-profits and not-for-profits occupied the upper tier.

The post-Katrina health-care landscape inherited this two-tiered system, but also cycles of reform that prioritized a bricks-and-mortar solution to the economic burden of uncompensated care, rather than proposing a new, more equitable system. At the same time, such a solution recapitulated a view held decades earlier. As one physician had put it, "If the Good Lord would give us the benefit of his abilities and warn us twenty-four hours in advance so the hospital could be evacuated . . . [He] could open the bowels of the earth and let Charity sink into the molten lava and slime and disappear so we could start all over. . . . That would be the best solution."[28]

The health-care void after Katrina provided a tabula rasa on which to fashion alternatives to the two-tiered system. Health-care critics presented Louisiana's system as aberrant, compared to other states where medical services for the insurance-poor and uninsured are more equally distributed between public and private sectors, and the poor more likely to use office-based physicians. Poverty could not explain this exceptionalism, because other states with high poverty rates do not exhibit Louisiana's inequitable pattern. Some critics evoked cultural practices, like that of New Orleans physicians who automatically orient nonpaying patients to Charity Hospital. Reformers' proposals rested on a behavioral economic rationale that, given choice, patients will choose better, private hospitals, although the considerable inpatient and outpatient use of Charity and University by Medicaid patients before Katrina troubles that assumption. Nevertheless, two movements clamored for change: Charity's own administrators, and a parallel group interested in primary health care.[29]

As the Charity system's economic survival depended on fluctuating oil and gas revenues, it periodically faced financial crises, aggravated by medical school competition for Charity patients, skimming off private-pay patients from the public sector, and deterioration of Charity's facilities. Economic survival of the charity hospitals came to depend on a funding mechanism, Medicaid Disproportionate Share Hospital payments (DSH),[30] which reimburses hospitals with large numbers of poor or low-income patients for costs incurred by uncompensated care. Louisiana's $1 billion DSH funds, far above the national average, helped multiply Charity Hospital's budget by a factor of five. Together with Medicaid, it accounted for 82 percent of the ten charity hospitals' total budget. Unsustainable and with little accountability attached, DSH monies allowed Louisiana to continue financing the charity system, thereby discouraging cheaper, more effective, ambulatory care. Its lower matching requirements also enabled Louisiana to divert money to road construction, education, and other non-health purposes.[31]

In 1989, the federal agency that certifies Medicaid and Medicare reimbursement threatened Charity with loss of accreditation unless improvements were made. In 1992, under pressure from the Joint Commission on Accreditation of Healthcare Organizations (JCAHO) to meet national standards, the state purchased a new hospital, hoping also to attract private patients.[32] The facility was named "University Hospital," and with Charity Hospital became the Medical Center of Louisiana at New Orleans (MCLNO). But in 1994 JCAHO stripped Charity Hospital of its accreditation, for life-safety and other code violations. The LHA then spent $1 million "without a logical financial basis" to design a new trauma center and critical-care tower, which were never built. Finally, patients were moved from overcrowded Charity to University Hospital, essentially displacing LSU's private-pay patients, who turned to the private sector. JCAHO eventually reaccredited Charity Hospital but in 2002 cited Charity and University hospitals, by then LSU-run, for endless cycles of repair and breakdown, safety problems, lack of patient privacy, and poor infection control. It recommended "strongly . . . seeking from the state a more modern facility."[33]

Plans for a six-hundred-bed replacement of the Charity facility were enthusiastically announced the following year: "After years of dreaming, the long sought hopes for a new state-of-the-art full-service health-care facility are going forward. We are 'on board a fast moving train!'" exclaimed Dwayne Thomas, CEO of the Medical Center of Louisiana, as he and a leadership corps representing the LSU Board of Supervisors, the Louisiana legislature, the LSU Health Sciences Center, and the LSU and Tulane schools of medicine met to present "an exciting action plan that will move toward actual construction, which consultants on the project visualize to be well underway toward completion in 2008, LSU Health Sciences Center's consultants recommended."[34] The January before Katrina, LSU provided a timeline for the new facility. The planning phase was to begin in late 2005 and construction in late 2007, with the completed, occupancy-ready facility projected for September 2010.[35]

Pre-Katrina, then, hospital-centered, bricks-and-mortar solutions prevailed over health system change, with reform practically identified with buildings. The JCAHO findings, to which proponents for replacing Charity would continuously return after Katrina, concern the structure and conditions of Charity and University facilities, rather than how care is accessed, provided,

or paid for. But LSU also hoped a new hospital would shed the stigma identified with Charity's structure and recoup insured patients lost in competition
with private hospitals.

Parallel to bricks-and-mortar proposals, a nascent movement emerged
around alternatives to the 150 primary and specialty care clinics physically
and administratively linked to Charity Hospital. Such facilities contradicted
the incipient national vision that primary care clinics should be community-
based "medical homes,"[36] a concept that today incorporates evidence-based
medicine, electronic records, managed care, and other principles. Community
health reformers founded the Partnership for Access to Healthcare (PATH) to
promote the principle of efficient and equitable health care through medical
homes. Thus, before Katrina hit, not only had LSU announced plans for a new
facility, but 13 primary care clinics dotted the city, according to the Louisiana
Public Health Institute. Charity Hospital, however, still provided the bulk of
care (80 percent), not only to the poor and to African Americans, but also for
acute and emergency care beyond the scope of small clinics.

In Katrina's Wake: Charity Hospital as a Boundary Object

In June 2006, LSU's Healthcare Services Division announced a Memorandum of Understanding (MOU) between the VA and LSU to explore "mutually-
beneficial consolidation" of the facilities, replacing their pre-Katrina ones.
Discussions were underway about how the two new hospitals could share
maintenance services, facility support, and clinical areas. The announcement
recalled the plan released in January 2005 for replacing Charity Hospital by
2012. But while the VA's financing seemed assured, Charity's was snarled in
disagreements with FEMA.[37]

That spring, advocates for re-opening Charity had requested the New Orleans City Council to seek the state attorney general's opinion on the legal-
ity of closing Charity. (He eventually ruled that if LSUHSD had closed Charity
Hospital, it should have been in compliance with law.)[38] Advocates then en-
rolled national health-rights attorneys and local attorneys to bring a class-
action suit on behalf of former Charity patients against Charity administrators, for illegal closure. Over the next two years, a coalition including the
Committee to Reopen Charity, Save Charity, and Doctors for Charity combined
their expertise, media knowledge, and organizing skills. Local television and
talk radio provided self-designated Charity Hospital Babies with a platform.
At public hearings required for the VA- and (later) LSU-proposed facilities,

advocates joined residents and business owners threatened with eminent-domain takeover.

The coalition sought an immediate response to health needs, through the reopening of the historically symbolic hospital. LSU countered that the new facility was "a done deal," complete with architectural plans. But over the next four years, LSU, the state, and FEMA would argue over the $1.5 billion LSU was seeking to finance Charity's replacement. An alternative meanwhile surfaced from a state legislature mandate, HCR 89, requesting the Foundation for Historical Louisiana (FHL) to assess conditions at the Charity facility and its potential for medical services. Although FHL's purview tends more towards preserving churches and Indian mounds, the foundation accepted its mission, and under a steadfast volunteer leadership raised the necessary funds and solicited bids from architectural firms. RMJM Hillier, one of the largest worldwide (it specializes in hospitals and designed the Louisiana Cancer Research Center in New Orleans), was hired through a bidding process. In August 2008, the firm released its controversial findings: Charity Hospital was structurally sound and capable of being transformed into a state-of-the-art modern facility. Furthermore, the report estimated construction costs significantly lower and time to rebuild far shorter than those the building of a new facility would entail. The report sparked debate about numerous assessments LSU had previously commissioned, which state legislators and FEMA criticized as institutionally self-interested and the consultants as non-independent.[39]

As both sides debated the report's conclusions, preservationists joined efforts to conserve Charity Hospital, given its status as a prime example of New Orleans Art Deco architecture. The National Trust for Historic Preservation named Charity Hospital and the Lower Mid-City neighborhood threatened with removal to make way for the medical campus to its list of "America's Eleven Most Endangered Historic Places." Thus, a new community of practice appropriated Charity Hospital not as a health-care issue per se, but as a preservationist concern.

Alongside preservationists, an overlapping community of practice emerged around social justice concerns regarding 250 homes and businesses threatened with eminent-domain takeover. In fact, fears of losing homes people had returned to rebuild after Katrina had been voiced as early as 2006, during the first neighborhood recovery plan commissioned by the City Council. Residents in Planning District 4, location of the Tulane-Gravier area slated for the VA and LSU hospitals, had voiced concerns about displacement and their Lower Mid-City neighborhood being torn down. During the next recovery planning process, the United New Orleans Plan (UNOP), the hospitals' effects on neighborhoods were discussed and alternative land use plans drawn up, but excluded from the final reports. Two years later, the master plan process drew anger from potentially affected residents of those Mid-City neighborhoods. They requested and obtained additional meetings, but not the desired public hearings. The Reopen Charity coalition, by this time including fifty local neighborhood associations and national organizations like the American Planners Association and National Trust for Historic Preservation, found itself translating health issues into the language of site selection, historic preservation, and social justice. Additional lawsuits and administrative appeals were filed to halt displacement and hospital construction.[40]

As each community of practice attributed its own meaning to the idea of Charity, the VA and LSU consciously reoriented their arguments for constructing the hospitals, developing economic rationales. With the hospitals slated for construction within a biosciences economic development district, they now termed the hospitals "the most important economic driver" in future New Orleans. For example, the VA's hospital project was alternately defined as meeting urgent health needs of veterans (albeit years after the storm) and, through the biomedical district "built upon the VAMC," as serving "as an economic engine that creates jobs for the entire city."[41]

The master plan further diverted attention from the hospital as a citywide public-health good. Like disaster monies, which funded projects but not overall recovery, the plan fragmented the city into planning districts, composed of smaller units for zoning. Despite this, the consultants retained to develop the plan, Goody-Clancy, maintained a larger vision. At the public meeting where they submitted the master plan draft to neighborhood leaders, the Goody-Clancy group's director christened the Charity Hospital the "open wound" of the master planning process.[42] In an unpublished memo[43] to the City Planning Commission during the citizen input period, Goody-Clancy expressed dissatisfaction with the city's lack of leadership on the hospital issue and communicated citizens' concerns about potential displacement of residents and blight to the neighborhoods the VA and LSU appropriated, as well as to the downtown Medical District, where Charity stood shuttered. After the master plan comment period, Goody-Clancy criticized LSU's plan for an academic teaching hospital on grounds that it disregarded the very principles laid out in the master plan itself, like urban density and a smaller footprint. The criticisms came too late. The City Council's inertia, Mayor Ray Nagin's behind-the-scene dealings,[44] and political and business support for a new medical facility had hastened the demise of Charity Hospital.

The Charity Hospital controversy and related VA hospital issues had garnered public attention for several reasons. On the one hand, they touched African American, poor, and disabled New Orleanians' fears of being left without a health safety net. Some whites disparaged and racialized these fears in public forums; professionals often labeled activists as ignorant and lacking expertise. On the other hand, because the hospital controversy could be reduced to bricks and mortar—something that everyone could fathom, unlike the complexity of health-care reform—the controversy was able to galvanize dissimilar groups. Bricks-and-mortar obfuscated labyrinthine arguments about health-care system financing and delivery laid out by critics of the two-tiered system. At the same time it transformed the meaning of Charity Hospital into a boundary object: an architectural treasure for preservationists; meaningful employment for health-care workers; crucial mental-health beds for mental-health consumers, families, and security-conscious police; a maternal, familiar medical home—albeit not the new kind—for people of color, the uninsured, the working poor, and marginalized citizens.

The more abstract health-care issue retreated before the damaging materiality of moving homes and businesses, transforming typical New Orleans neighborhoods into suburban-like sprawl.[45] FEMA symbolically laid the first brick when it finally agreed, in arbitration, to pay the state $475 million for Charity Hospital's damage, a leap from their initial $23 million offer— although for years LSU's ability to finance the hospital would continue to be disputed by legislators and media.[46]

On the Ground: Health-Care Rationalization and Primary Care Experimentality

The parallel, primary care experimentality emerged through ad-hoc storefronts and clinics established in Katrina's wake by faith-based groups, community activists, retired health professionals, and local medical and allied health schools, staff and students. These often grassroots efforts incorporated actors from the pre-Katrina primary care movement and some former Charity personnel, such as the Charity nurses who opened a clinic out of the Lower Ninth home of one of them. The Louisiana Public Health Institute (LPHI), collaborating with the Centers for Disease Control (CDC), assembled this loose network to develop a plan for rebuilding Southeast Louisiana's health-care delivery system. They produced a "Framework for a Healthier Greater New Orleans," whose objectives were incorporated by the first recovery planning body, the Bring New Orleans Back Commission (BNOB). BNOB's Health Committee proposed up-to-date technology to end Louisiana's charitable populism and a new facility to replace the building that embodied it, as well as a comprehensive system and private-public provider mix. U.S. health secretary Michael Leavitt then unsuccessfully attempted to impose a model for Louisiana health-care reform, through the state's Health Redesign Commission, that would divert Medicaid money into vouchers for private insurance. But the primary care experimentality evolved into a hundred-clinic network, bolstered by a Primary Care Access and Stabilization Grant (PCASG) from special disaster funds and manpower resources. Cited by President Obama's health secretary, Kathleen Sebelius, as a model for the nation, warranting replication, the community-based clinic network became the poster child for post-Katrina health care.[47]

The experimentality in primary care presents an image of enterprising success by contrast to the supposedly backward gaze of Charity proponents. The Charity Hospital controversy and the primary care network both illustrate how disaster funds—albeit from different pots—were harnessed to fast-forward plans envisioned before Katrina. But, whereas the primary care network had to prove sustainability, FEMA finally provided LSU and the state with funds towards building a new hospital, without requiring that LSU prove its ability to finance construction and hospital operations.[48]

Disaster capitalism, frequently evoked in recovery analyses, hardly explains these contradictory examples. The Bush administration did try to privatize healthcare through vouchers; private hospitals indeed bought up shuttered facilities after Katrina; and the blow to Louisiana's unique, public charity hospital system set off a domino reaction, including closure of Baton Rouge's charity hospital. The state began withdrawing from health-care delivery in favor of privatization and undermined the state civil service. But privatization glosses over important dynamics informed by historical contingencies that shaped the debates in New Orleans.[49]

But the growth-machine model of social vulnerability to disasters is also necessary to understanding the cycles of reform that led to abandoning Charity Hospital for a new, not-yet-built facility. Sociologists define "growth machines" as "a process that is built and set in motion by persons who focus on profit and 'progress' . . . but that has no internal brakes and no sensors to take note of the damage it is doing as it churns along."[50] The model fits environmentally risky, humanly engineered projects, like the Mississippi River Gulf Outlet that facilitated the flooding of New Orleans. Long before Katrina, cycles of reform harnessed the bricks-and-mortar efforts that carried over after the disaster and culminated in a future LSU-VA-Biosciences District, thereby transforming health-care delivery into an economic engine. Katrina thus fast-forwarded growth-machine processes already underway, against which embodied situational identity fueled an opposition movement. Like the means by which communities of practice expanded the Charity Hospital "object" through reinterpretations, Charity Hospital Babies engaged not in culturalist interpretations of the hospital, but rather in the practical work of culture inextricably entwined with material conditions and vital needs.

Conclusion

Reformers and patients produced two major narratives of health-care reform after Katrina. The successful primary care experimentality provided New Orleans with a nationally acclaimed narrative of redemption, not from disaster only, but from corruption, collective ineptitude, widespread poverty, and supposedly antiquated pre-Katrina health-care ideals. The work of Charity Hospital Babies produced a different sort of narrative, which demanded recognition of the very conditions that redemption symbolically overcame. The movement to reopen Charity, and the variations on that movement, projected a public version of what transpired at a micro-level, in the neighborhood planning process. There, citizens often bypassed the constraints set by zoning codes and other planning principles to express desire, to produce wish lists and critical analyses. They needed, as one planner put it, to tell their saga, the story of their neighborhood and its uniqueness. By creating a public narrative, Charity Hospital Babies met a similar need, without losing sight of the material conditions that drove them to resist LSU's plans.[51]

Viola Green, with whom I opened this essay, was evacuated from New Orleans to Thibodaux, then to Detroit. For weeks, she prayed "to remember the [phone] number of someone who wasn't in the water." Her daughter-in-law had stayed behind with her own mother, a chemotherapy patient confined to a wheelchair, afraid of leaving. Both were eventually rescued from their rooftop. In Detroit, a Red Cross worker finally located one of Viola's brothers and booked her on a flight to join him in North Louisiana. Frances, meanwhile, had evacuated to Houston with her youngest son and his teenage friends, becoming the "Mama" in the apartment complex where a volunteer agency placed them. Back home in New Orleans, she squatted in the abandoned houses of her damaged neighborhood, struggling with drugs and depression, until a women's residence took her in. Meanwhile, Viola suffered multiple strokes while living in a FEMA trailer on her brother's property. She died in 2007. Neither she nor Frances ever learned what happened to Mr. Bailey. Mr. Jackson passed away shortly after the storm.

Those social conditions of life, health, and death have barely changed for many New Orleanians. How people in post-Katrina New Orleans perceive their health today remains dramatically split along race lines, with blacks almost three times as likely as whites to fear health services might not be available to them. Blacks are still more likely to report the diseases of poverty: hypertension, respiratory problems including asthma, and diabetes.[52] And one-fourth of New Orleanians report that emergency departments are their only health-care source. Addressing these conditions in the future will largely depend far more on the fate of the federal Patient Protection and Affordable Care Act of 2010 than on the turbulent struggle for health reform in Katrina's recovery.

Notes

1. Research for this essay was funded by an Agence National de la Recherche grant, ANR-07-BLAN-0008-22. Heartfelt thanks to Sue Makiesky Barrow, Helen Regis, Eveline Thevenard, Samuel Bordreuil, Stéphane Tonnelat and participants in the Tulane-EHESS Katrina workshop for their incisive comments on earlier versions of this paper. All names are aliases, except for public figures and interviewees who gave permission to identify them. Descriptions are from field notes and interviews, unless otherwise noted.

2. Hundreds of New Orleanians remained stranded on the elevated Interstate 10, near Causeway Exit, under a burning sun.

3. Rebecca Solnit, *A Paradise Built in Hell: The Extraordinary Communities That Arise in Disasters* (New York: Viking, 2009); quotation in Stephen Lukes, "Questions About Power: Lessons from the Louisiana Hurricane," Vilhelm Aubert Memorial Lecture at the Institutt for Samfunnsforskning in Oslo, Norway, September 22, 2005, published online in "Understanding Katrina: Perspectives in the Social Sciences," Social Science Research Council, on June 11, 2006, understanding katrina.ssrc.org/Lukes/.

4. Franco Basaglia, "The Disease and Its Double," in *Psychiatry Inside Out: Selected Writings of Franco Basaglia,* ed. N. Scheper-Hughes and A. M. Lovell (New York: Columbia University Press, 1987); Anthony Oliver-Smith and Susanna M. Hoffman, "Why Anthropologists Should Study Disasters," in *Catastrophe and Culture. The Anthropology of Disaster,* ed. Susanna M. Hoffman and Anthony Oliver-Smith (Santa Fe, N.M.: SAR Press, 2001): 3–22; Nancy Scheper-Hughes, "Katrina: The Disaster and Its Double," *Anthropology Today* 21, no. 6 (2005): 1–2.

5. Two-thirds of the six thousand displaced physicians were from the central New Orleans parishes (www.unc.edu/news/archives/sep05/); William R. Freudenburg, Robert B. Gramling, Shirley Laska, and Kai Erikson, *Catastrophe in the Making: The Engineering of Katrina and the Disasters of Tomorrow* (Washington D.C.: Island Press, 2009).

6. Massachusetts provides a well-known example with its innovative, near-universal health-insurance plan, one template for the 2010 federal Affordable Care Act.

7. Geoffrey Bowker and Susan Leigh Star, *Sorting Things Out: Classification and Its Consequences* (Cambridge Mass.: MIT Press, 2000).

8. Personal communication to author, November 18, 2008.

9. Quotation in Susan Larson, "Lt. General Russell Honoré Offers Lessons in Getting Ready for Disasters," *New Orleans Times-Picayune,* May 6, 2009, blog.nola.com/susanlarson/2009/05/lt_general_russel_honore_offer.html (downloaded January 2, 2010).

10. Don Smithberg, "Message from Don Smithberg, CEO Health Care Services Division. All HSCSD employees and patients displaced by Hurricane's Katrina and Rita, please call the hotline number to check in," Baton Rouge: Louisiana State University Health Sciences Center homepage, 2005.

11. Deposition of James P. Moises, February 6 and 16, 2009, for *LeBlanc v. Thomas,* Civil District Court Parish of Orleans. See also James Gill, "LSU Won't let the Facts Get in Hospital's Way," *Times-Picayune,* May 6, 2009, blog.nola.com/jamesgill/2009/05/lsu_wont_let_facts_get_in_hosp.html (downloaded May 7, 2009).

12. The National Environmental Policy Act and Section 105 of the National Historic Preservation Act require resource surveys and public review processes before federally funded development projects, such as building a VA facility, can be undertaken.

13. Kevin U. Stephens Sr. et al., "Excess Mortality in the Aftermath of Hurricane Katrina: A Preliminary Report," *Disaster Medicine and Public Health Preparedness* 1 (2007): 15–20.

14. Anne M. Lovell, "Debating Life after Disaster: Charity Hospital Babies and Bioscientific Futures in Post-Katrina New Orleans," *Medical Anthropology Quarterly* 25, no. 2 (2011): 254–77; Martha C. Ward, *Poor Women, Powerful Men: America's Great Experiment in Family Planning* (Boulder, Colo.: Westview Press, 1986.)

15. Jonathan Roberts and Robert Durant Jr., *A History of the Charity Hospitals of Louisiana* (Lewiston, N.Y.: Edwin Mellen Press, 2010); Harriet A. Washington, "Apology Shines Light on Racial Schism in Medicine," *New York Times,* July 29, 2008.

16. Roberts and Durant, *A History of the Charity Hospitals of Louisiana;* John Salvaggio, *New Orleans' Charity Hospital: A Story of Physicians, Politics, and Poverty* (Baton Rouge: Louisiana State University Press, 1992).

17. Roberts and Durant, *A History of the Charity Hospitals of Louisiana;* Kent Germany, *New Orleans after the Promises: Poverty, Citizenship, and the Search for the Great Society* (Athens: University of Georgia Press, 2007).

18. This cultural racism is more consonant with racist interpretations "non-whites" held about discriminatory abandonment associated with Katrina (see Diamond essay in this volume) than with color-blindness evident in other aspects of contemporary New Orleans.

19. The entire quotation reads: "It ain't fair. . . . Honest boys with good records come out of LSU and can't get into that medical school. You gotta have a lot of money. That medical stuff comes high. And then they only let in a certain number anyway. Louisiana needs doctors. We're a-going to fix that—a free medical school, and there won't be a place to equal it" (Harnett T. Kane, *Huey Long's*

Louisiana Hayride: The American Rehearsal for Dictatorship, 1928–1940 [Baton Rouge: Pelican Publishing, 1941], 222–23). At the time, only the private Tulane Medical School trained doctors in New Orleans, mostly middle- or upper-class "whites," many from out of state. This set off the proverbial animosity between LSU-Charity and Tulane University.

20. Salvaggio, *New Orleans' Charity Hospital;* Daughter of Charity (no date), "Daughters of Charity New Orleans," www.thedcno.org/about.html.

21. Roberts and Durant, *A History of the Charity Hospitals of Louisiana,* 281; Ward, *Poor Women, Powerful Men.*

22. Trauma, HIV/AIDS, substance use, and psychiatric patients included nonpoor and insured people.

23. Yi Fu Tuan, "Language and the Making of Place: a Narrative-Descriptive Approach," *Annals of the Association of American Geographers* 81 (1991): 684–96; Denise Lawrence-Zuniga, "Cosmologies of Bungalow Preservation: Identity, Lifestyle and Civic Virtue," *City and Society* 22, no. 2 (2010): 211–36; Anne M. Lovell, "'The City Is My Mother': Narratives of Homelessness and Schizophrenia," *American Anthropologist* 99, no. 2 (1997): 355–68; during reform cycles, the maternal Charity figure gave way to demeaning, whore-like imagery, as if she had shed her maternal protection. One state official described the Tulane-LSU medical school competition thusly: "Charity Hospital is like a whore in Venice playing both sides of the canal—Tulane and LSU medical schools. . . . This is its biggest problem!" (Salvaggio, *New Orleans' Charity Hospital,* 227). A devoted Charity physician told physician-author Salvaggio: "Charity, like other major urban public hospitals, resembles an old whore who [sic] people use and cast aside; state legislators simply say, 'You're no longer my fancy,' and that's the end of it" (306).

24. On neighborhood identity in New Orleans, see the oral histories from the Neighborhood Story Project: www.neighborhoodstoryproject.org

25. Roberts and Durant, *A History of the Charity Hospitals of Louisiana,* 308; Lovell, "Debating Life after Disaster."

26. Karen Gadblois, cited in Anne M. Lovell, S. Bordreuil, and V. Adams, eds., "Public Policy and Publics in Post-Katrina New Orleans: How Critical Topics Circulate and Shape Recovery Policy," *Kroeber Anthropological Society* 100, no. 1 (2011): 118–19.

27. Quotation in Bring New Orleans Back Commission, *Report and Recommendations to the Commission* (New Orleans: Bring New Orleans Back Health and Social Services Committee, 2006), 29; Public Affairs Research Council of Louisiana, "Reform the Health Care Reform Process, PAR Says," June 1, 2010, www.dynasite.net/s3web/1002087/preview.cfm/parpublications/commentar iesandletters/100008.

28. Quotation in Salvaggio, *New Orleans' Charity Hospital,* 218.

29. On efforts to reform the two-tiered system, see Anne M. Lovell, "Who Cares about Care? Health Care Rationalization and the Demise of a Public Hospital after Katrina," *Metropolitiques,* www.metropolitiques.eu/?lang=en; R. Rudowitz, D. Rowland, and A. Shartzer, "Health Care in New Orleans before and after Hurricane Katrina," *Health Affairs* 25, no. 5 (2006): w393–w496; PriceWaterhouseCoopers, *Report on Louisiana Healthcare Delivery and Financing System,* 2006, lra.louisiana.gov/assets/docs/searchable/reports/PwChealthcarereport42706l.pdf; Roberts and Durant, *A History of the Charity Hospitals of Louisiana,* 227.

30. DSH funds are funneled through state Medicaid programs to hospitals the state designates as serving a "disproportionate share" of low-income or uninsured patients. The funds are in addi-

tion to regular payments for inpatient care of Medicaid beneficiaries but in Louisiana substituted for them.

31. Interview with David Hood, PAR health policy expert, December 1, 2010; Public Affairs Research Council of Louisiana, *Realigning Charity Health Care and Medical Education in Louisiana,* May 2007, www.parlouisiana.com/s3web/1002087/docs/Publications/Realigning_Charity_May_2007.pdf.

32. The 1982 needs-assessment report stated that "the facility must be improved in view of the fact that CHNO [Charity Hospital at New Orleans] must compete with private care facilities for paying customers—Medicaid, Medicare, insured. The crowded facility with open wards will not have a chance to compete for revenues if it is not improved. . . . [A]n improved facility is the first step in an effort to better meet the needs of the two medical schools and to attract the paying clientele" (Sunbelt Research Corporation, cited in Anonymous, 1982).

33. Roberts and Durant, *A History of the Charity Hospitals of Louisiana,* 49–54, 55–56.

34. Medical Center of Louisiana, "New Medical Center of Louisiana: A Major Step Toward Reality," *This Week at Medical Center of Louisiana,* Baton Rouge, July 7–13, 2003.

35. Roberts and Durant, *A History of the Charity Hospitals of Louisiana,* 56.

36. On medical homes, see the special issue of *Health Affairs* 20, no. 5 (May 2009).

37. "Proposed Joint LSU VA Facility," *Highlights,* LSU Health Care Services Division, April 30, 2007, www.lsuhospitals.org/documents/LSU-VA_Highlights.

38. Opinion number 07–0169, June 18, 2007.

39. RMJM Hillier, *Medical Center of New Orleans: Feasibility Study* (Baton Rouge: Foundation for a Historic Louisiana, 2008), www.newhospital.org/plans/Historical%20Louisiana%20plan%202.pdf; Bill Barrow, "Charity Hospital Debate Turns on Distrust of Expert Assessments," *Times-Picayune,* December 3, 2009, www.nola.com/politics/index.ssf/2009/12/charity_hospital_debate_turns.html (downloaded December 4, 2009).

40. *Tulane/Gravier Neighborhood Planning District Four Rebuilding Plan* (New Orleans, undated); interview by author with Keith Scarmuzza, architect and urban planner, May 26, 2009.

41. E-mail from Maya Gorman, Louisiana Solutions (agent for the city), to Nathalie Carlis, Louisiana Disaster Recovery Unit, February 10, 2009, thelensnola.org/wpcontent/uploads/2010/09/GormantoCarlis.Correspondence0001.pdf.

42. Goody-Clancy presentation of the master plan draft to neighborhood leaders, Dryades YMCA, March 21, 2008.

43. "Medical District: Assessment of Current Planning and Zoning Issues," draft for review and comment from David Dixon and Mary Means to Yolanda Rodriguez (City Planning Commission), November 5, 2008.

44. Nagin signed a Memorandum of Understanding giving the state the right to proceed with eminent-domain takeover of parts of Mid-City, without going through any public process. The VA did not need similar proceedings because the federal government has jurisdiction in eminent domain and need not comply with city zoning laws.

45. Horizontal parking lots were built on land where LSU planned to build an academic hospital.

46. Associated Press, "Public Hospital Board Scraps Federal Finance Idea," July 7, 2011.

47. Louisiana Public Health Institute, *Louisiana Public Health Institute Annual Report, 2004–2005* (New Orleans: Louisiana Public Health Institute, 2005), 15; Mary A. Clark, "Rebuilding the

Past: Health Care Reform in Post-Katrina Louisiana," *Journal of Health Politics, Policy and Law* 35, no. 5 (2010): 743–69; Chelsea Ledue, "HHS Waiver to Boost New Orleans Community Clinics, Access to Care," *Healthcare Finance News,* September 24, 2010.

48. Bill Barrow, "LSU Officials Tell Skeptical Legislators University Medical Center Can Be Elite Regional Hospital," *Times-Picayune,* May 2, 2011.

49. Naomi Klein, *The Shock Doctrine. The Rise of Disaster Capitalism* (New York: Picador, 2007).

50. William R. Freudenburg et al., *Catastrophe in the Making: The Engineering of Katrina and the Disasters of Tomorrow* (Washington, D.C.: Island Press, 2009), 10.

51. Margaret R. Somers, "The Narrative Construction of Identity: A Relational and Network Approach," *Theory and Society* 23 (1994): 605–49; Lovell, "'The City Is My Mother.'"

52. Henry K. Kaiser Family Foundation, *New Orleans Five Years after the Storm: A New Disaster amid Recovery* (August 2010), 62.

The Political Economy of Invisibility in Twenty-First-Century New Orleans

Security, Hospitality, and the Post-Disaster City

THOMAS JESSEN ADAMS

It has become a truism in American political discourse that Hurricane Katrina eroded a physical and social facade that screened out the poverty, racism, and disrepair that plagued a large swath of New Orleans and the Gulf South. Nationally, political commentators across the spectrum have credited the 2005 storm and the bungled response of federal, state, and local governments as the constitutive moment in broader public opinion's decisive shift against the presidential administration of George W. Bush. As *New York Times* columnist Frank Rich put it in his book *The Greatest Story Ever Sold,* "the true Katrina narrative was just too powerful to be papered over by White House fictions."[1] In the San Francisco *Chronicle,* Marc Sandalow opined, "Hurricane Katrina's winds ripped away barriers that kept one city's poor out of sight, and for most people, out of mind."[2] *Washington Post* communist and MSNBC commentator Eugene Robinson chimed in: "it took a conspiracy of woe to create the human conditions that Hurricane Katrina unmasked."[3] Even President Bush, in his nationally televised September 15, 2005, speech in New Orleans's Jackson Square, noted how surprised Americans were to witness the suddenly visible "deep, persistent poverty" of the Gulf South and New Orleans in particular.[4]

For its part, scholarly treatment has echoed the political commentary in its emphasis on narratives of Katrina as a moment of unmasking and visibility. Cultural theorist Henry Giroux has argued, "Katrina broke through the visual blackout of poverty and the pernicious ideology of color-blindness to reveal the government's role in fostering the dire conditions of largely poor African-Americans."[5] Aimee Berger and Kate Cochran pointed out that Katrina helped lay bare the way "the South functions in the nation's social imaginary to contain and make invisible racism and poverty."[6] Historian J. Mark Souther noted in a special issue of the *Journal of American History* devoted to Katrina that the image of starving, ill-clothed, destitute, black and brown bodies huddled together on freeway overpasses and crumbling rooftops "laid bare the persisting relevance of race and poverty."[7] In the same issue, Alecia Long, in a trenchant critique of the ongoing destruction of public housing in the city, noted that "the crime that the poor of New Orleans are most guilty of is making themselves so damn visible" (803).

More recently and popularly, the writer Dave Eggers, in his bestselling account of Abdulrahman Zeitoun's months-long unconstitutional detention in the aftermath of Katrina, and a similar subplot in HBO's ode to New Orleans culture, *Treme,* have begun to draw attention to the so-called Katrina Disappeared—the hundreds of inmates in Orleans Parish Prison who were lost for months on end in a truly Kafkaesque maze of indefinite detention and suspension of constitutional rights.[8] The growing interest in the stories that came out of Orleans Parish Prison and rural parish jails as exemplified by the plight of Zeitoun and the fictional brother of Khandi Alexander's character, Ladonna Batiste-Williams in *Treme,* as well as some surges of national interest in police brutality during the storm as demonstrated by the cover-ups and prosecutions of the Danziger Bridge and West Bank police shootings, has further reinforced the motifs of visibility and invisibility, exposure and unmasking, as central to the way Americans have understood the storm.[9] Whether it is the sudden undeniability of poverty and racism on the nation's collective television screen during the storm, the growing fascination with one of the biggest constitutional violations in recent American history in the case of the Katrina disappeared, or the plaudits and outrage that have greeted A. C. Thompson's truly excellent reporting on the details of police and white vigilante violence during the storm, Katrina and its aftermath have been seen as a singular moment where the facade of modern American society was peeled away and the poverty, racism, and legal and extra-legal violence that it entails were revealed for all the world to see.[10]

Indeed, the motif of exposure, of uncovering and making visible the nefarious aspects of what lies beneath the standard picture of American society at the beginning of the twentieth century, has given the image of post-Katrina New Orleans an almost noirish character in the eyes of Americans at large. In the same way that critics like Mike Davis have argued that mid-century Los Angeles functioned as a rotten and decayed backdrop for revealing the corruption at the heart of the American Dream, New Orleans has come to embody the ultimate location where a new rot at the core of the United States is being revealed.[11] At the same time, New Orleans and the Gulf South, through the motif of their uncovered façade, are functioning much like the broader, "backward" South has always functioned for liberal American opinion. Nearly a half-century ago, historian C. Vann Woodward commented that the South has operated as a space that enlightened northerners can look at while patting themselves on the back for their supposed progressiveness. As Woodward wrote, "North and South have used each other, or various images and stereotypes of each other, for many purposes... not only to define their identity and to say what they are *not,* but to escape in fantasy from what they *are.*"[12] Indeed, much of the metaphor of exposure and visibility that permeates the nation's understanding of Katrina is more about the nation itself than those in New Orleans and the Gulf South. Segregation, racism, poverty, and white vigilante violence are what happen below the Mason-Dixon line, or alternately, though all too similarly, Katrina showed us that New Orleans is America *writ large.* In either narrative, though, the subject remains not New Orleanians or their city, but the whole of the nation.

The problem with the narrative of visibility and invisibility, as expressed in virtually all media, scholarly, and popular outlets, is its disconnect from political economy—from the very forces that not only produced the invisibility to begin with, but continue to do so, more so now than ever. Invisibility is in fact the constitutive cultural characteristic of the political economy of New Orleans, more so in the aftermath of Katrina than ever before. While the storm may have exposed the so-called invisible poor and the destitute living conditions of America's most poverty stricken and disrepaired major city, it in fact furthered the connection between the political economy of invisibility and economic dislocation.

Arguably, "invisible commodities" are more central to New Orleans than they are to any other city in the United States, save perhaps Las Vegas. As goods that are produced, owned, sold, and consumed, academic disciplines such as history, economics, sociology, and critical social theory have rarely reckoned with commodities that do not take tangible form. Yet, in post-Katrina New Orleans, the city, broadly construed, has little to sell but intangible objects. The vast majority of economic activity in the city can be understood as attaching economic value to social meaning. In New Orleans, the commodities the city has to sell hinge on the ability of residents, new migrants, and consumers to place an economic value on a certain kind of experience. Consumers *pay* for authenticity, desire, service, luxury, tradition, and security. Furthermore, beyond direct consumption, the immense profits made in the post-Katrina real estate market are directly tied to the invisible labor that produces New

Orleans's "last-frontier appeal" of "authentic charms"—the constitutive commodity of the city's growing gentrification.[13] The ways in which two of these
commodities—security and service—are produced, managed, and sold in the
recent history of New Orleans tell us much about the political, racial, and economic character of the post-disaster city.

New Orleans and its economy have not always been dependent on selling
such incorporeal and invisible commodities. In the last half of the twentieth
century a series of disasters—technological, economic, and environmental—
served to unmoor the city from virtually any connection to the so-called "traditional" anchors of blue-collar employment and economic growth. First, the
relatively simultaneous growth of Houston as the financial center of the oil
industry and Miami as the economic and cultural gateway to Latin America
supplanted New Orleans's earlier positions as both the financial and administrative center of the Gulf South petroleum industry and the nation's leading trading partner with Central and South America. Later, the city's relatively
slow adoption of containerization shifted its port, for more than a century
one of the world's busiest and America's most geographically important, into
the second tier in the hierarchy of the nation's shipping. Later in the 1980s,
the downturn in oil prices wiped out a brief boom in extraction employment,
oil service, and petroleum finance. Indeed, the 2010 Deepwater Horizon oil
spill off the Gulf Coast will likely serve to further diminish local petroleum
employment given national cries for drilling moratoria while diminishing
one of the region's last vestiges of physical commodity production—the Gulf
Coast seafood industry.[14]

For at least the last two decades, though, all of these industries—oil, shipping, Latin American import/export, and seafood have taken a back seat to the
region's biggest employer and largest moneymaker—the hospitality industry.
By the time Marc Morial took office as mayor in 1994, the city's transformation from financial center of petroleum extraction and key global shipping
node to an economy overwhelmingly based on tourism and service was almost complete. Forced to compete for convention and leisure dollars with the
relatively sanitized spaces of Las Vegas, San Diego, Orlando, Orange County,
and metropolitan Phoenix, the Morial administration made a concerted effort to sell the city and its attractions' safety and physical security. In 1995,
Morial formed the Tourism Industry Leadership Task Force (TILTF) for the
express purpose of stemming purported declines in visitors as a result of the
city's reputation, noting that there was "a trend for groups considering New
Orleans to question whether they should come because of national news

about crime here."[15] According to the TILTF, "there has never really been in the past any noted tendency for groups to question their plans to come here," and the industry is "now worried that we would be facing a Miami like situation with a significant fall off in visitation."[16] The task force was referencing a wave of crime against tourists in early 1990s Miami, including a series of murders and carjackings that led to the cancellation of hundreds of conventions and a marked downturn in the region's own tourism-based economy.[17] By the mid-1990s, New Orleans was fully aware that its biggest problem in luring more tourists lay in the city's perception among out-of-town visitors as being unsafe. Indeed, according to a Gallup poll commissioned by the New Orleans Convention and Visitor's Bureau, "safety and crime rate was the number one response for least positive aspect of conducting tours and conventions" in the city.[18]

For a city supposedly as inefficient and bureaucratically backward as New Orleans, Morial's TILTF operated like a Swiss train regarding concerns about safety and security. Tourist complaints of crime and "street harassment" (typically synonymous with aggressive panhandling) were met with immediate replies and a host of vouchers and coupons. At the extreme end of this response is the story of Daniel Renaud, a Toronto man visiting the city for the first time in 2000 who wrote to Morial to complain about the crime he experienced. As he told Morial, "On February 23rd . . . I was assaulted by a group of youths on the corner of Loyola and Canal. I was punched in the jaw and fell to the ground where I was repeatedly kicked in the throat and face. . . . Not a person offered any assistance nor did anyone attempt to stop the attack. Seconds after the incident a police car passed by but did not stop to offer assistance despite my friends frantically waving at the officers in the car."[19]

What seems remarkable about Renaud's experience, and that of other tourists who were, or claim to be, victims of crime in New Orleans is the speed and generosity with which Morial's office attempted to placate them. By March 31, barely two weeks after receiving Renaud's letter, the mayor's office had sent Renaud a voucher on Air Canada for a return trip and three nights at the Sheraton, as well as coupons for a riverboat cruise and various meals, and a special dinner party hosted by the city's Council of International Visitors. While Morial and the city were exceedingly generous in the case of Renaud, offering vouchers, coupons, and even a "key to the city" to disgruntled tourists who experienced crime was the norm. Lest the city's actions in these cases be understood as altruism, the mayor's press secretary noted how it "was a great opportunity for the mayor and the city to get some good p.r. . . .

[T]he story might appear in other markets. This has the potential to be a win/ win situation."[20] The Morial administration's interest in transforming the city's reputation for visitors and tourists was unprecedented at the time. Previous mayors like Sidney Barthelemy and Dutch Morial had seemingly paid little attention and less money to combating New Orleans's image as crime-ridden in the minds of potential tourists.

Thus, for at least a decade before Katrina hit, the relationship between crime, national reputation, and the city's economy was well understood by policy makers. Katrina and the fabricated and false images of a city overwhelmed by racial disorder and its descent into a kind of Hobbesian, state-of-nature chaos represented a turning point not in a concern with crime, but in the overwhelming entrance of security into the marketplace. Cities and urban planners have always concerned themselves with crime as a broad social problem; what Katrina helped usher in was not a concern with crime per se, but the growing need to purchase security, often for its own sake. Even the most regressive and ineffective anticrime measures and policies, from branding and mutilation in the premodern city to mandatory minimum sentencing today, were explicitly designed to either prevent crime, change the criminal, or remove him or her from society.[21] Security on the other hand is directed not toward the criminal or the crime, but the marketplace, cash nexus, and consumer. It is a feeling or experience that occurs under certain social conditions. As it became a truism that the aftermath of Katrina entailed social breakdown, lawlessness, rampant theft, and racialized violence, a massive market was created to both sell this feeling and experience and to use its existence or lack thereof as a justification for the transformation of social space in the city.

In post-Katrina New Orleans, security as a commodity has been instrumental in the almost wholesale transformation of the city's once large concentration of public housing into private, mixed-income development. The 1980s and 1990s saw a shift in national urban policy that resulted from the coalescence of the policy concerns of the new-right, free-market revanchism, the middle-class rediscovery of the city, and the culture of poverty, which generally removed large populations of the poor and racial and ethnic minorities from real estate that was rapidly multiplying in value.[22] What is of particular interest in this new urban-planning era is the way in which the concept of "defensible space" has not simply been incorporated into the demolition of public housing, in physicality and as a right, but has in fact become the primary reason in the case of New Orleans for the destruction of the city's hous-

ing projects and the production of security as a constitutive commodity in the
city's redevelopment.

"Defensible space," in its classic formulation by architect Oscar Newman,
argued that the massive housing projects of the industrial North and Midwest created a series of physical spaces outside the watchful eye of produc-
tive community members.[23] Criminals were thus more likely to gravitate to
places such as corridors and stairwells where residents and police would not
witness them engaging in crimes. At the same time, housing developments
that featured residents who were more invested in the community, usually
through paths to ownership or ownership itself, would tend to "defend" their
spaces from criminality. As a concept, "defensible space" was a constituent
aspect of the Clinton-era HOPE (Homeownership and Opportunity for People
Everywhere) VI program that led to the wholesale demolition of public housing in many American cities from the mid-1990s through the present. In New
Orleans, though, the logic of "defensible space" gained particular traction after Katrina. While in most cities it was part of new design concepts meant to
alleviate crime and the so-called social pathology of the urban poor, in New
Orleans it also became a justification for itself and the profits of those who
provided it.

In 2008, HUD, in conjunction with the Housing Authority of New Orleans
(HANO), began demolishing the old "Big Four" public housing projects in the
city, Lafitte, B. W. Cooper, C. J. Peete, and St. Bernard. Curiously, despite these
buildings having no spatial or architectural resemblance to the high-rise
model of northeastern and midwestern public housing—the physical layouts
that Newman originally singled out as so problematic—a HUD spokeswoman
argued that New Orleanians deserved much better than the physical layout of
the crumbling "buildings which basically warehoused the poor" and spread
crime and social ills.[24] The bastardized version of "defensible space" employed by HANO and its associated private developers argued that the public housing stock in New Orleans must be demolished not because the old
projects were indefensible—indeed, under the traditional theory, a few minor
landscaping adjustments could have made them model physical spaces for
Newman's theory—but because residents of both subsidized housing and the
city as a whole needed to feel secure for the city's *economic* and *commercial*
future. To that end, "defensible space" has been employed by HANO and, in
particular, private developers like Columbia Residential, which redeveloped
the old St. Bernard Projects into the mixed-income Columbia Parc using de-
fensible-space rhetoric as the main selling point for the new community.[25]

Similarly, at the old Magnolia Projects (C. J. Peete) in Central City, McCormack Baron Salazar (MBS), the St. Louis–based developers who won the contract for the new Harmony Oaks development, argued in their 2008 press release to commemorate the new community's groundbreaking that the new design would revitalize all of the Central City neighborhood by offering safe, secure, and high-quality housing to residents of a variety of incomes.[26]

Companies like Columbia Residential and MBS have found New Orleans to be an especially lucrative market for their design concepts. HANO, the Louisiana Industrial Development Board, and the Louisiana Office of Community Development have outlaid over $170 million in public financing for the new Harmony Oaks development. Most telling, though, is where MBS received its private funding. With little working capital on hand, MBS has almost fully funded its construction costs in partnership with Goldman Sachs. Initially, Harmony Oaks will offer 193 units of public housing, 144 units of subsidized (which for the developer and investors comes out to the same as market rate), and 123 market-rate units. Additionally, MBS was allowed to construct 50 houses, dubbed "Harmony Homes," which it can sell at market rate.[27] Over time, MBS's requirement to allocate public housing will decline, leaving it with essentially unfettered ownership of twenty square blocks of prime real estate, conveniently located halfway between New Orleans's two major centers of wealth, Uptown and the Downtown/French Quarter areas.

Contrasting the fate of Magnolia with that of the old Florida Projects in the Upper Ninth Ward makes both the logic of Goldman's investment and the impulse toward security clear. The Florida Projects, isolated from key economic parts of the city and bounded by the Industrial Canal and a group of abandoned warehouses, were abandoned after Katrina and have attracted little of the comparable interest that greeted the Magnolia redevelopment and the possible future redevelopment of Iberville. MBS and Goldman have seen the mixed-income redevelopment of Magnolia as an opportunity and not simply as a land grab; indeed, if that were the case, developers would be knocking down HANO's door to flip Florida into mixed-income units. Rather, the city, state, and federal governments turned to partners like MBS and Goldman to purchase security in developmentally important areas like the Central City corridor. In turn, Harmony Oaks can now sell that security back to the city and its residents at a tidy profit. The perceived demand for safe and secure housing, amplified and enabled after Katrina, not for low-income residents themselves, but for the future of New Orleans's economic development, opened up possibilities for developers to supply new, "defensible space" to the city and its residents. In the end this is a process whereby capital, in its literal form in the case of Goldman's investment in Magnolia, is able to latch onto a cultural demand for public safety and the feeling of security and turn it into a vendible commodity.

While the political economy of invisibility has manifested itself in the selling of subjective security back to the city in the form of newly privatized housing, invisibility as a constitutive commodity in the post-Katrina city is also evident in the way labor is organized in New Orleans's largest industry—hospitality. In a 2009 interview to promote his HBO show, *Treme*, David Simon expressed, albeit uncritically, the constitutive aspects of the city's political economy. As Simon told the *New York Times*, "lots of American places used to make things. Detroit used to make cars. Baltimore used to make steel and ships. New Orleans still makes something. It makes moments."[28] Moments, experiences, feelings are in fact exactly the commodities made in New Orleans and, as a direct corollary, the places where value and profit are produced. And like cars in Detroit or steel in Baltimore, they are made by people under specific political, labor, cultural, and social conditions and in contingent historical contexts. Indeed, Simon's "moments" are not static examples of some imagined authentic culture, but produced, managed, and sold commodities. Their ability to be successfully sold and commoditized are utterly dependent on the ability of the people whose labor produces them to be rendered socially and politically invisible.

Hospitality, like the other products the city increasingly sells, depends upon the social erasure and invisibility of the workers who produce it and their quotidian labors. Specifically, within a broad economy based on hospitality, managers and owners needed to organize their workforce in order to convince customers that their employees are not individuals and humans, but products and commodities. Indeed, people ranging from academic economists to McDonald's franchisees have understood that, the more service and hospitality workers are seen as human beings performing a given service for another human being, the less efficient and profitable a given service-providing business would be.

Theodore Levitt, a longtime Harvard business scholar, clearly appreciated the larger economic stakes in invisibility as it relates to service and hospitality labor. As he put it in his enormously influential 1972 *Harvard Business Review* article, "The Production Line Approach to Service," "service thinks humanistically, and that explains its failures."[29] Levitt, who less than a decade later would become somewhat famous for his coinage of the term "globalization," was attempting to formulate a management philosophy to solve one of the most vexing problems in the modern economy—how to increase profitability, efficiency, and labor control when American business increasingly sold intangible services rather than physical commodities. The problem, according to Levitt, was that far too often, business owners, managers, and customers saw in the sale and purchase of a given service, not the buying and selling of commodities like cleanliness, efficiency, and comfort, but the buying and selling of human beings such as the individual janitors, health aides, waitresses, maids, line cooks, security guards, and busboys who provided these services. Levitt argued that this alienation was an obstacle to greater profitability, as efficiency, labor-saving strategies, and automation were less likely to occur when managers and customers viewed service workers as individuals rather than products and commodities.

For a business to be economically successful, a kind of cultural and psychological distance was required between consumer and server. While a typical physical commodity by nature of its visible, physical existence was able to screen out the exploitation and alienation inherent in its production, the intangible commodity central to profitability in an economy based around hospitality has no such luxury. Indeed, the consumption of services provided by individual human beings all too often inspired a feeling—in manager, consumer, and server alike—of pre-capitalist labor relations, of slavery, feudalism, and European social deference. As Levitt wrote, "The concept of service evokes, from the opaque recesses of the mind, time-worn images of personal ministration and attendance. . . . [I]t carries historical connotations of obedience, subordination, and subjugation. . . . [P]eople serve because they are compelled to, as in slavery."[30] For Levitt, countless managers, and business owners, the key to rationalized service operation was the removal of the overly and overtly human aspect of service work, be it through turning a perceived identity into a vendible commodity or, much more commonly, rendering an entire class of workers virtually invisible, both culturally and economically.

In certain regards, Levitt's comparison of hospitality to slavery was apt. Service work, and particularly the low-wage and highly capitalized personal service industries that grew in such dramatic fashion across the country after World War II and are so central to the American economy shared a central paradox with slavery. If slavery meant the buying and selling of human beings as commodities, then the service economy, with its similar selling of individuals rather than tangible, physical commodities, engendered somewhat closely related problems in a liberal, market-based society dependent on the self-ownership of free individuals. If liberal freedom grounded in possessive individualism is based on the assumption that no individual can own another, then the buying and selling of human beings in any form represents a profound cultural problem. Indeed, service workers at times seemed to be selling more than their labor, but rather their entire selves.

Also like Atlantic slavery, service work and the production of service workers as commodities, like the production of captured Africans and their offspring as commodities, were deeply intertwined with race. Without revisiting decades-old historical debates about the primacy of race in the making of Atlantic World slavery, we can at least generalize that race and slave status operated dialectically as ever more Africans were able to be enslaved because they were conceived of as a different race, while at the same time slave status helped to further fix African lineage as a separate race. And so it has been with service work. The labor of African Americans and Latino immigrants was more easily sold as commodities because they were racialized as not-white. Consumers and managers alike were able to ignore the troubling feeling that when one purchased a service, one was not simply purchasing labor, but purchasing a body in large part because of the racial classification of those bodies. At the same time the segmentation of African Americans and Latinos into service occupations reified a racially divided society that helped further entire strata of the population's invisibility.

Levitt's understanding of service is a particularly apt place to start when analyzing the racial/labor geography of New Orleans, especially after Hurricane Katrina produced a wholesale reorganization of the city's demography and economy. The New Orleans hospitality industry includes over 2,800 restaurants, bars, and hotels. Collectively, these establishments employ more than 110,000 of the region's estimated 520,000 workers.[31] Over the last twenty years, employment in hospitality has grown at a greater rate than any other sector in New Orleans. Between 1990 and the eve of Katrina, hospitality jobs nearly doubled across the New Orleans metropolitan region while the rest of the area's employment remained virtually stagnant. Since Katrina, this trend has been exacerbated, as hospitality jobs have nearly returned to their 2005 levels while non-hospitality jobs across the region remain at only 75 percent of their pre-Katrina levels.[32]

The character of the workers in the industry has changed since Katrina as well. Most significantly, the age of hospitality workers has gotten decidedly older since Katrina, with more than 22 percent of workers over the age of forty-five, compared to just 13 percent before the storm.[33] These numbers suggest the inadequacy of the broad cultural understanding of hospitality work. In the popular narrative, hospitality work is considered a way station for the young and upwardly mobile who are seen as on their way to another, more lucrative and stable career. At the same time, within traditional understandings of what constitutes working-class work, labor that does not realize itself in a physical commodity, like service and hospitality work, is seen to be emasculating, female, and generally outside the category of socially valued labor.[34]

Also significantly, the industry has become less African American and decidedly more Latino and Asian. While African Americans still make up at least 40 percent of the hospitality workforce in metropolitan New Orleans, that number is down from 45 percent before Katrina. Asian Americans, especially Vietnamese, displaced from their primary regional node of employment in the shrimping and fishing industries, and Latinos, a plurality of whom are of Honduran descent, have picked up the difference, each doubling their numbers in the industry since Katrina and Rita (*Behind the Kitchen Door,* 7).

Across the board, the pay of these jobs lags well behind that of other regional employers. With the exception of managers, concierges, chefs, and back-of-the-house line supervisors, no position in the New Orleans hospitality industry averages over ten dollars per hour. Indeed, of the more than 110,000 workers in the industry, fully 84 percent make less than ten dollars an hour, including tips. More than 10 percent of workers in the entire industry make less than minimum wage, including tips (14).

Unsurprisingly, benefits in the industry are virtually nonexistent. More than 85 percent of workers report that their employer does not provide health insurance. Even those who do have the option can rarely afford it. As one twenty-year veteran of New Orleans bartending put it, "even if I had the option to get health benefits, I wouldn't be making enough to afford the health benefits" (14). Or, as a male busser who moved to New Orleans right after the storm said, "My benefit is me working and getting tips" (15).

Similarly, advancement within New Orleans hospitality is exceedingly rare. A longtime waitress put it like this: "your job is your set job where you work. . . . [I]f you a doorman, that's what you're gonna be, you ain't gonna be nothing else. If you come in there and want to be a bartender, you always going to be a bartender, always. . . . [T]hem managers is keeping their manager positions, they ain't going nowhere" (16).

Legal violations are also, unsurprisingly, incredibly common in the industry. Stealing tips and lack of overtime pay are common. As one longtime server said, "How they do you is they will work you extra hours, they will work you overtime, but what they do is in every establishment they separate you, so you got one place there, one here and they split you up and they give you a different time card for every place that you work at, and each one is separate so you never really work over time. . . . [T]hen, as far as tip-outs of, see, however many workers you might have, two people over here, three people over there working daiquiris and you got you manager sitting on his ass right here. . . . [A]t the end of the night the only person who can touch the money is the manager, and he counts all the money and splits the tips between everybody that's working and himself. He gets a tip portion too . . . so you ain't get nothing" (19). Indeed, wage theft like that described by the above server is endemic, especially in restaurants that cater to tourism in the French Quarter.

Racial segmentation is particularly dramatic in the way employment is structured in the hospitality industry and contributes significantly to the racialized invisibility of the labor force. Over 78 percent of the industry's white workers work in front-of-the-house jobs, while nearly 68 percent of African Americans work in back-of-the-house employment (43). Latino workers have become similarly ensconced in back-of-the-house work following an increased migration to South Louisiana in the years after Katrina. As one long-time bartender put it, "Definitely there are Hispanics working back there, and it's interesting because they are the ones in the back, and they are the ones doing the hard work . . . but there wasn't a single Mexican or Hispanic server. All caucasian, all college kids, all white, uppity on top of that. Everyone in the back . . . to them it's almost like a favor you know, be grateful we even give you this job kind of thing so they kind of took a lot of abuse in like verbal, and racial slurs. Usually people in the background are minorities" (42). Or, as a woman who's been a line cook for three decades succinctly put it, "the majority of workers in the front are white. Everybody in the kitchen, ain't no white, all black" (42).

A particularly astute bartender at an elite French Quarter restaurant described the front-of-the-house, back-of-the-house divide in the terms of both gender and race: "In New Orleans, which has been a predominately black city, there's been this since colonial times, well-established sort of totem pole of who works in the service industry. In fine dining there's a preponderance of males working the floor, you notice if you go into those restaurants its male dominated and also European American males. So the people who work their way up the totem pole more often are male, more often are white" (43). The location of hospitality workers within the divide between front and back positions, or what might be called the line between visibility and invisibility, largely determines wages, mobility, and workplace safety. Nearly a fourth of the city's front-of-the-house workers report making what they deem to be a living wage, while only 2 percent of those who work behind the scenes can claim to make a living wage (42). Similarly, the predominantly African American population of back-of-the-house staff experiences decidedly higher rates of job-related injuries, including burns, cuts, and toxic-chemical exposure. In terms of mobility between positions, or even to higher-end jobs as line supervisors, many African Americans reported a "glass ceiling" and little opportunity for advancement. As one African American woman succinctly put it, "you can be as smart as this book right here but they won't hire people like me for certain positions" (46).

The fact that the New Orleans hospitality industry, like virtually every other hospitality industry in the nation, is engulfed by ongoing racial discrimination is hardly surprising. The discrimination, though, is usually not the direct result of racist hiring practices, but a general desire on the part of the city's industry to screen out from public view the largely black and brown bodies that work in the industry and pass through the same physical spaces as customers, tourists, and the city's growing population of middle-class and wealthy post-Katrina migrants. The service and hospitality workers who are essential to the profitability of New Orleans's largest industry are, like security, essential to the post-disaster city's economic growth. Their labors, and indeed, oftentimes, their selves are also like security, unseen and invisible as something bought and sold. Too often in New Orleans accounts of Katrina and its aftermath emphasize exposure and uncovering, bringing to physical visibility the existence of previously unseen atrocities. In these tellings, though, not only does agency rest with the viewer—the appalled American citizen witnessing New Orleans poverty for the first time on television in late August 2005, or even the valiant investigative reporter delving into the inner workings of a broken penal system—but the political economy that in fact produces security as the justification for the wholesale destruction of public housing and the hidden labor of hospitality workers as the backbone of the city's largest industry and the profit margins of its wealthiest residents and investors remains, in a word, invisible.

Notes

1. Frank Rich, *The Greatest Story Ever Sold: The Decline and Fall of Truth, From 9/11 to Katrina* (New York: Penguin, 2006), 201.

2. Marc Sandalow, "Katrina Thrusts Race and Poverty onto National Stage," *San Francisco Chronicle,* September 23, 2005, 1.

3. Eugene Robinson, *Disintegration: The Splintering of Black America* (New York: Doubleday, 2010), 118.

4. George W. Bush, "Address to the Nation on Hurricane Katrina Recovery from New Orleans, Louisiana," September 15, 2005, www.hurricanekatrinanews.org/Bush.html (accessed May 10, 2010).

5. Henry A. Giroux, "Violence, Katrina, and the Biopolitics of Disposability," *Theory, Culture, and Society* 24 (2007): 309.

6. Aimee Berger and Kate Cochran, "Covering (Up?) Katrina: Discursive Ambivalence in Coverage of Hurricane Katrina," *CEA Forum* 36, no. 1 (Winter 2007).

7. J. Mark Souther, "The Disneyfication of New Orleans: The French Quarter as Façade in a Divided City," *Journal of American History* 94, no. 3 (December 2007): 804.

8. Dave Eggers, *Zeitoun* (San Francisco: McSweeney's, 2009).

9. See, for instance, Campbell Robertson, "Charges Filed in Katrina Inquiry," February 24, 2010, www.nytimes.com/2010/02/25/us/25orleans.html (accessed May 12, 2010); Robertson, "6 Are Charged in Post-Katrina Shootings," *New York Times,* July 13, 2010, www.nytimes.com/2010/07/14/us/14justice.html (accessed July 14, 2010).

10. See A. C. Thompson, "Katrina's Hidden Race War," *The Nation,* January 5, 2009; A. C. Thompson, Brendan McCarthy, and Laura Maggi, "New Orleans Police Department Shootings after Katrina under Scrutiny," December 13, 2009, *New Orleans Times-Picayune*; and Rebecca Solnit, *A Paradise Built in Hell: The Extraordinary Communities That Arise in Disaster* (New York: Viking, 2009). It should be noted that this violence was local common knowledge long before Thompson and fellow journalists like Solnit, McCarthy, and Maggi began publishing their investigative journalism in 2009.

11. Mike Davis, *City of Quartz: Excavating the Future in Los Angeles* (New York: Verso, 1990), 36–46.

12. C. Vann Woodward, *American Counterpoint: Slavery and Racism in the North-South Dialogue* (New York: Little, Brown, 1964), 6.

13. Sara Ruffin Costello, "Love among the Ruins," *New York Times Style Magazine,* October 6, 2013, 67. Costello's article is part of an emerging genre extolling the authenticity of New Orleans and its supposedly unique culture. This genre itself plays a key role in screening out the labor that imagines the city as a playground of unmediated cultural consumption, an imagining that allows for the city and particularly, its real estate, to be sold at immense profits.

14. For the best scholarly treatment of the development of the New Orleans hospitality industry, as well as a concise account of the city's larger economic history since World War II, see J. Mark Souther, *New Orleans on Parade: Tourism and the Transformation of the Crescent City* (Athens: University of Georgia Press, 2006), especially 159–220. Also see Kent Germany, *New Orleans after the Promises: Poverty, Citizenship, and the Search for the Great Society* (Athens: University of Georgia Press, 2007).

15. Memorandum, Jeanne Nathan to Mayor Marc H. Morial and Police Chief Richard Pennington, "Meeting of Tourism Industry with Chief Pennington and Mayor," February 21, 1995, New Orleans City Archives, Records of Mayor Marc H. Morial (hereafter MHM), Tourism Files, Box 4, Tourism Crime Meetings.

16. Ibid.

17. On Miami's crime wave and the attendant drop in tourism, see Judy Holcomb and Abraham Pizam, "Do Incidents of Theft at Tourist Destinations Have a Negative Effect on Tourists' Decisions to Travel to Affected Destinations," in *Tourism, Security and Safety: From Theory to Practice,* ed. Yoel Mansfield and Abraham Pizam (Boston: Elsevier, 2005), 108–24.

18. Gallup Organization, on behalf of New Orleans Metropolitan Convention and Visitors Bureau, "New Orleans as a Destination: Results of a Survey of Meeting Planners and Tour Operators," 1997, Tourism Files, MHM, Box 3, Crime/Tourism.

19. Daniel Renaud to Mayor Morial, n.d. (received by Mayor's Office on March 15, 2000), MHM, Tourism Files, Box 5, Complaints-Crime.

20. Dean Shapiro to Rhonda Spears, March 31, 2000, MHM, Tourism Files, Box 5, Complaints-Crime.

21. See, for instance, Roger Lane, *Policing the City: Boston, 1822–1885* (Cambridge, Mass.: Harvard University Press, 1977); Michel Foucault, *Discipline and Punish: The Birth of the Prison* (London: Penguin Books, 1977); Peter Linebaugh, *The London Hanged: Crime and Civil Society in the Eighteenth Century* (New York: Verso, 1991).

22. For a generalized critique of these trends, see Neil Brenner and Nik Theodore, eds., *Spaces of Neoliberalism: Urban Restructuring in North America and Western Europe* (Oxford, U.K.: Blackwell, 2002); Adolph Reed Jr., *Stirrings in the Jug: Black Politics in the Post-Segregation Era* (Minneapolis: University of Minnesota Press, 1999), 163–96.

23. Oscar Newman, *Defensible Space: Crime Prevention through Urban Design* (New York: Macmillan, 1972).

24. Donna White, quoted in Katy Reckdahl, "New Designs Hope to Avoid Past Problems in Public Housing Complexes," *Times-Picayune,* May 22, 2009, 1.

25. See www.hano.org/index.php?q=node/49 (accessed August 23, 2010).

26. Press release, "McCormack Baron Salazar, KAI Design & Build Begin Construction on New Orleans Neighborhood Redevelopment Project," January 9, 2008, www.mccormackbaron.com/CJ -Peete_FINALGroundBreakingPressRelease.doc (accessed August 23, 2010).

27. See www.hano.org/index.php?q=node/50 (accessed August 23, 2010).

28. Wyatt Mason, "The HBO Auteur," *New York Times Magazine,* March 17, 2010, www.nytimes. com/2010/03/21/magazine/21simon-t.html?pagewanted=all (accessed May 18, 2010).

29. Theodore Levitt, "The Production-Line Approach to Service," *Harvard Business Review,* September–October 1972, 41–52.

30. Levitt, "The Production-Line Approach to Services," 43.

31. Restaurant Opportunity Center of New Orleans, Restaurant Opportunities Center United, and the New Orleans Restaurant Industry Coalition, *Behind the Kitchen Door: Inequality, Instability, and Opportunity in the Greater New Orleans Restaurant Industry* (New Orleans: Restaurant Opportunity Center, 2010), 4. (Hereafter *Behind the Kitchen Door.*)

32. Ibid., 5.

33. Ibid., 7–8.

34. See, Thomas Jessen Adams, "The Servicing of America: Service Work and Political Economy in Postwar Los Angeles," PhD diss., University of Chicago, 2009, 2–20.

Faith, Hip-Hop, and Charity

Brass-Band Morphology in Post-Katrina New Orleans

BRUCE BOYD RAEBURN

During the five years since Hurricane Katrina, African American brass-band musicians and scholars in New Orleans have undergone some revealing experiences. These include the storm's influence on attitudes among brass-band musicians about the nature of the tradition, essentially a matter of faith in response to an uncertain future; coping strategies of market-driven forces related to supply and demand (especially with regard to hip-hop and cultural tourism), as well as community-based aesthetic imperatives pertaining to the survival of heritage; and the consequences of interventions by scholars, volunteer organizations, and corporations, all of which qualify as charity in the broadest sense of "to think well of others," as well as conforming to the narrower definition of "aid for the destitute."[1] These issues show how Hurricane Katrina affected preexisting conditions in New Orleans, as well as how the disaster created new challenges and opportunities. Rather than being fatalistic about outcomes, I choose to present an optimistic vision, admitting full well that I may be wrong. The current economic forecast for most New Orleans musicians is certainly not a rosy one, but that was true before Hurricane Katrina. Yet music is intrinsic to lifestyle in New Orleans, and consequently money is not always the primary criterion of value, as it is for the American popular music industry in general.

For more than a century, brass bands have figured prominently in the New Orleans jazz environment, a singular phenomenon which distinguishes the city's musical infrastructure from other locales where jazz is performed throughout the United States and abroad. From about 1900 until the 1960s, most jazz-related brass-band activity was tied primarily to the city's festival traditions, ranging from city-wide celebrations such as Carnival to more circumscribed rituals such as "second-line" parades, church dedications, and brass-band funerals reflecting the community life of predominantly African American neighborhoods such as Tremé, the Ninth Ward, Central City, and Gert Town/Hollygrove.[2] The founding of Preservation Hall in 1961 and the emergence of Harold Dejan's Olympia Brass Band, which experimented openly with repertoire and instrumentation, in 1962 acted to shift the balance increasingly toward cultural tourism, creating a bifurcated market situation that supplements the original economic imperative of service to neighborhood communities with a plethora of assignments made possible by a burgeoning cultural tourism industry.[3] Thus, in order to make ends meet, African American brass band musicians especially have had to incorporate performances for tourists at conventions, in the streets, and at nightclubs with an annual schedule of community "second lines" and funerals sponsored by social aid and pleasure clubs and benevolent associations.[4]

While the expansion of the market created new economic opportunities for these musicians, it made them increasingly vulnerable to any forces or events that would disrupt the cultural tourism and community-based "second lines" on which their livelihoods depended. Thus, by simultaneously wiping out or otherwise disrupting both tourism and the community life of several of the most crucial neighborhoods that had sustained "second-line" parades for more than a century, Hurricane Katrina hit New Orleans brass bands where they lived, effectively effacing their market and jeopardizing the survival of the city's jazz tradition.[5] Yet the city's most popular and innovative brass bands at the time—Hot 8, Soul Rebels, and Rebirth—returned early as opportunity permitted, based largely on faith in the resilience of the market and the tradition, and they were generally successful in re-situating themselves in a market that still remains in recovery mode. They had brand power going for them. Younger bands, such as TBC (To Be Continued), Baby Boyz, and Young Fellaz, were either just starting out when Katrina hit (TBC was formed in 2002) or were organized afterward. Baby Boyz and Young Fellaz were both founded in 2007, which means that the recovery market is all they have ever known. Nevertheless, the faith in the viability of the tradition that motivated the older bands to return to the city after Katrina is present in the younger generation. Rosalie Cohn writes of Sean Roberts (TBC), Glenn Hall III (Baby Boyz), and Ashton Hines (Young Fellaz): "The three musicians possess a youthful optimism that is paired with the complete confidence that playing music is what they were born to do. . . . Open-minded and unrestricted, TBC, Baby Boyz, and the Young Fellaz function like sponges, searching for inspiration, and learning from virtually everything they encounter."[6]

Hurricane Katrina intensified and clarified a number of issues pertaining to cultural dynamics within the brass-band community. How would catastrophe affect the divisions between traditionalists and modernists that existed among (and within) brass bands prior to the storm? How would the bands reconfigure themselves to meet the shifting mandates of a market in recovery mode? Would music provide a way back into the culture, a means to preserve it, or would touring become an escape route? These questions can be explored by examining the commercial, social, and artistic imperatives that currently drive the fortunes of black brass bands in New Orleans. But equally heroic and important for the future is the proactive involvement of scholars who care about the tradition.

Obtaining accurate statistical information about the displacement or return of Crescent City musicians is still problematical, although the situation has improved during the last five years. Local projections that vary according to the constituency may be less reliable than outsider perspectives that focus on specific transactions, such as the provision of instruments for "2,400 professional musicians" from New Orleans during Phase One of the charitable organization Music Rising's strategic plan in 2006.[7] By comparison, the nonprofit Sweet Home New Orleans estimates that there were 4,500 musical practitioners, including musicians, Mardi Gras Indians, and social aid and pleasure club members in New Orleans prior to Katrina.[8] According to its August 2008 report, Sweet Home New Orleans found that by June 2007 two-thirds of these musicians and collateral tradition bearers had returned to the city but nearly half were experiencing difficulty in finding housing and work, which paid less when jobs were forthcoming. The 2008 report estimated that, by August 2008, 75 percent of the musicians had returned to New Orleans, yet problems related to housing and gainful employment persisted.

The welfare of musicians has been intrinsic to the idea of recovery in New Orleans, and the restoration of cultural practices associated with African American brass bands, especially "second lines" and "jazz funerals," has been used repeatedly in the media as a barometer of the city's recovery efforts.[9] Charity initiatives promulgated through concerted and sometimes coordinated private-sector engagement, rather than the anticipated governmental response, has been the key to "bringing New Orleans back," despite the exploitation of that phrase by municipal task forces that generated little more than publicity.[10] The actions of non-profit organizations such as the National Association of Recording Arts and Sciences' (NARAS, the Grammy Foundation) Musicares initiatives, Music Rising, the Musicians Clinic, Sweet Home

New Orleans, and Habitat for Humanity, along with numerous voluntary church and school groups from across the nation, compensated in the minds of many New Orleanians for the perceived inability of local, state, and federal governments to effectively manage recovery. It was largely this voluntarism from the private sector that sustained hope for the future after the disaster, and much of it was focused specifically on the welfare of musicians. A hastily organized collaboration between Music Rising and Musicares was responsible for restoring instruments to New Orleans musicians in 2006 and 2007.[11] In addition to attending to the health-care needs of local musicians, after Katrina, the Musicians' Clinic paid for free "after hours" shows at Snug Harbor in the Faubourg Marigny, providing supplemental income to professionals in 2006–8 while they waited for the market to rebound.[12] Although Habitat for Humanity's Musicians Village was conceived in December 2005 to address the housing needs of New Orleans musicians without regard to stylistic preferences, many African American brass-band musicians were denied access to the program because their credit histories did not meet the eligibility requirements (a situation that has subsequently been addressed through a credit remediation program implemented by Habitat for Humanity).[13] In August 2008, the Roots Music Gathering of the Cutting Edge Music Business Conference held a session in which representatives of the New Orleans Jazz and Heritage Foundation, NARAS, the Musicians' Clinic, Sweet Home New Orleans, Tipitina's Foundation, Habitat for Humanity, Broadcast Music Incorporated (BMI), and the leader of the Tremé Brass Band, Bennie Jones, discussed potential collaborations and sharing of resources in meeting the needs of musicians still coping with dislocations resulting from Hurricane Katrina. This level of coordinated interest and action had not existed prior to Katrina.[14]

Katrina also attracted unprecedented scholarly and journalistic interest in the New Orleans brass band tradition, with several instances of "embedded" scholars working collaboratively with bands facing an uncertain future. Case studies by ethnomusicologists Jeremy Tolliver of Bethel University and Matt Sakakeeny of Columbia University (now at Tulane), along with the networking activities of Stella Baty Landis (formerly at Tulane), provide insight on the post-Katrina responses of musicians in three of the city's top brass bands—the Soul Rebels, Rebirth, and Hot 8, respectively. In addition, Dr. Michael White, a clarinetist and jazz scholar at Xavier University, has expressed himself in various interviews and writings, representing the tribulations of a musician who lost everything except his talent to the storm.

If any doubts remain about the resiliency of New Orleans musicians or concerns about their commitment to their cultural heritage, these case studies should dispel them. Musicians who were forced into exile by Hurricane Katrina were willing to travel vast distances at their own expense to perform back home. Despite the interest of the younger generation in melding hip-hop with brass-band styles (as in "bounce"), musicians have put their faith in the city's cultural heritage as the most effective means to revitalize a population paralyzed by despair. They are generating hope through the power of music, despite the display of "swagger" and related hip-hop tropes that belie it.[15] In his thesis, Tolliver quotes Samuel Winston of *Gambit Weekly* in April 2006 regarding the Soul Rebels commuting to New Orleans from their respective exiles in Houston and Baton Rouge for a weekly engagement at a bar on Magazine Street: "Despite what seems like an extremely impractical set of circumstances for the band to continue at Le Bon Temps, they insist on it being necessary."[16] In New Orleans, music is always a necessity. The composer W. T. Francis said as much in 1890, and it still rings true.[17]

Lumar LeBlanc, the band's snare drummer, felt that regaining contact with the local audience was important, even though its promotional strategy prior to Katrina had relied increasingly on national touring: "We've committed ourselves to our fans, and we'll do it as long as we physically can."[18] One might interpret this imperative as commitment to the marching band tradition, but in the case of the Soul Rebels, some qualifications are in order. Well before Katrina, the band was moving away from street culture in order to market itself more effectively as a hip-hop act. Shortly after its founding in the early 1990s it released two contradictory publicity photographs—one in suits and fedoras, the other in berets and camouflage. In retrospect, it is not difficult to see which side won. For the cover of the August 2006 issue of *Off-Beat* magazine, the band insisted on not being portrayed with brass instruments for fear of typecasting, reiterating the title of their 1998 CD, "No More Parades."[19] Considering the dominance of hip-hop artists in the popular music market, this strategy makes sense, but it does not preclude commitment to restoration of endangered brass-band traditions, especially in times of crisis. During his audition, which he passed because the band learned that he was gutting and painting flooded houses, Tolliver's desire to find vernacular "authenticity" may have influenced his first impression of the Soul Rebels as an amalgam of funk, rhythm and blues, and hip-hop "with an unmistakable New Orleans brass band feel." But he certainly did not mistake the renewed sense of agency and urgency that he witnessed as he tracked the band's activities. He noted that "everyone in the group had committed themselves to bringing New Orleans back, and they saw their music as being the best tool for that task."[20] Before Katrina the Soul Rebels were looking for a way out of the community-based brass-band world, with its street-oriented jazz funerals and "second lines." Now they were working their way back in, but on their own terms.

Hurricane Katrina forced a reevaluation of the brass-band heritage for all the bands that returned or commuted; musicians who lost all their worldly possessions were left to ponder the value of tradition and to assess what it meant to them. Catharsis led to some interesting surprises. For more than three decades there has been a rift within the city's brass-band community between "modern" and "traditional" approaches to performance practices and repertoire. Dr. Michael White has commented on this phenomenon:

> During the middle and late seventies, several new younger brass bands emerged to fill in the community void as traditional groups on the street became rarer and rarer. The ranks of younger players were filled with musicians who rarely interacted with older players and had little, if any, contact with or 'apprenticeship' in the authentic traditional New Orleans style. . . . Characteristic of many younger groups, which grew to dominate the streets by the early eighties, were unison section playing, 'riff tunes,' simplified harmonies, reduced numbers of keys, faster and more prominent rhythms, and smaller groups of about eight musicians. Instrumental shifts were also seen: The clarinet virtually disappeared, and the tuba and drums developed a more dominant role with rhythm and blues overtones. The overall sense of pride, professionalism, and seriousness gradually gave way to street clothes, baseball caps, and tennis shoes. Mixing all the musical elements around them—reggae, rhythm and blues, bebop, free jazz, high school band music, Mardi Gras Indian chants, current radio hits, and television themes—the younger brass bands brought a freshness and creative excitement to the streets that had been missing in the older traditional bands. Commercialism and the generation gap were the underlying causes of the older bands' dwindling repertoires, stagnation of styles, and a lack of new songs in the traditional groups.[21]

White lived a block away from the London Avenue canal breach and lost everything he owned due to the flooding occurring after Hurricane Katrina, except for the instruments he took when he fled. He evacuated to Houston but commuted regularly to New Orleans to perform and eventually to resume teaching at Xavier University, where he lived in a FEMA trailer. In October 2005, a reporter for the *New Orleans Times-Picayune* went along as he attempted to salvage what he could from the debris. "At this point," White admitted, "I'm trying to figure out if I can be salvaged. I tried very hard to picture what this would be like, but you can't begin to imagine. The hard part is that there's a lot of history here that can't be replaced. It's all gone." In reflecting on the past, White invoked the "jazz funeral," in which the spirit of the deceased is "cut loose" to enjoy a better life in the great beyond. Death is followed by rebirth. "I have to keep remembering that," he said, "That's what gives us the courage to carry on."[22]

Yet before Katrina, White had already begun to temper his staunch and seemingly exclusive commitment to traditionalism. As a recipient of an artist-in-residence fellowship at Studio in the Woods in 2004, he had expanded his musical horizons to include hip-hop, modern jazz, rhythm and blues, and Caribbean music, resulting in two dozen compositions that appeared on his *Dancing in the Sky* CD. Katrina seems to have reinvigorated both his traditionalism and his interest in experimentation, evident not only in an intensification in his playing but also in his willingness to work as a mentor with younger, "new wave" brass bands such as Hot 8.

In the months leading up to Hurricane Katrina, the word on the street was that Hot 8 had replaced Rebirth as the favorite among social aid and pleasure clubs, largely because of its creative rearrangement of such popular hits as Marvin Gaye's "Sexual Healing" in the quirky, idiomatic New Orleans brass-band style, keeping the song's hooks intact while introducing greater rhythmic and dynamic intensity. Yet despite its success, the band had already experienced a degree of misfortune prior to Katrina that was astounding. The Hot 8 was organized by brass bassist Bennie Pete in 1995. Between 1996 and 2004 the band lost two members as the result of murder: trumpeter Jacob Johnson was found shot "execution style" in his home, and trombone player Joseph "Shotgun" Williams was killed by the police during a traffic stop. In December 2006 the trend continued with the murder of snare drummer Dinneral Shavers by a teenaged assailant attempting to kill his stepson. The previous April, trumpeter Terrell Batiste had been hit by a car while he was attempting to change a flat tire in Atlanta, where the Red Cross had provided an apartment—he lost both legs but continued to perform with the Hot 8. Pete was coping with morale problems that would have paralyzed most bands, but his response was instead to seek new avenues for growth and discovery. He asked Dr. Michael White to collaborate in exposing his band members (who had studied music at Southern University in New Orleans) to the traditional brass band heritage, a subject not offered in schools.[23]

For White and Pete, drawing on the symbolic potency of a century-old continuum of brass-band jazz performance was a strategic initiative designed to sustain hope—it was a coping mechanism as well as a mingling of artistic perspectives. Apropos of White's remark about the brass-band funeral, Pete told a reporter for NPR at Shavers's funeral: "We feelin' that dirge. We expressin' that dirge to the dead. We feel in our mind he could see this some kind of way. . . . I brought my best for him on this day."[24] For a population enervated and dispossessed by disaster, the ability to feel, to get beyond the numbness, is a powerful antidote to despair. The weekly sessions held at Sounds Café illustrated how hope could enhance and redefine "hipness" in a post-Katrina environment. When queried about the traditional repertoire they were performing with White in February 2007, such songs as "Margie," "Bye and Bye," and "The Saints" (which is considered by some to be the ultimate New Orleans cliché), Hot 8 trumpeter Raymond Williams explained, "We got to play these songs; this is who we are."[25] Before Katrina, that statement would have been unthinkable. The trauma wrought by the storm blew away what now seemed to be artificial boundaries dividing the brass-band community and refocused attention on respect for the tradition as a whole—conceived as a holistic continuum in which every musician could aspire to a place of honor.

Hurricane Katrina helped a number of displaced bands to achieve greater national exposure. Hot 8 had a primarily local following until their appearance in Spike Lee's HBO documentary, *When the Levees Broke: A Requiem in Four Acts,* in 2006. According to one reporter, the result was that "a new legion of fans caught onto the band's mix of traditional marching music, hip hop, and R&B."[26] Did increased national visibility and an awareness of the heavy promotional machinery that would soon be dedicated to "jump starting" the city's cultural tourism industry factor into the band's reassessment of traditionalism? A more plausible explanation is the mediation of musicologist Stella Baty Landis. In 2007 she was a visiting instructor in Tulane University's Music Department with an interest in documenting the Hot 8, and she was also the owner of Sounds Café. In an interview with Howard Reich, Landis stated: "What we really need is to make [the artists] who are here, and who feel so strongly about the city, feel comfortable about staying here." In making her space available for the Michael White–Hot 8 collaborations, she became proactive in setting an agenda for sustaining the tradition and was not particularly concerned about scholarly detachment or stylistic boundaries. Yet even Landis was not certain that New Orleans's traditional culture would survive: "I feel that the city that existed pre-Katrina is done, and that its culture is not going to suddenly reappear, no matter how hard we fight for it," she said. "My hope is very long-term and itself kind of abstract, but it's that the kernel of New Orleans, that essence of New Orleans that is deep in our soul and thick in the air, will give rise to something, eventually, newly wonderful."[27]

In his dissertation, Matthew Sakakeeny explores the ways in which brass-band musicians are reinventing tradition.[28] Sakakeeny filmed funerals in 2006 and 2007, focusing on how bands such as the New Birth and Rebirth were adapting to the post-Katrina environment. These bands draw their membership and entourage primarily from Tremé, a neighborhood just north of the French Quarter that experienced minimal flooding but has nevertheless had to cope with perennial crime, poverty, deteriorating housing stock, and police repression. Well before Katrina, the eradication of an oak-lined public space along Claiborne Avenue to make way for an interstate highway in the 1960s and the expropriation of space to build Armstrong Park (intended as a playground for tourists) in the 1970s served as ominous messages from city officials that Tremé and its rich "second-line" traditions were considered to be expendable. The interstate overpass at Claiborne Avenue and Dumaine Street therefore holds a special significance—it symbolizes the disrespect Tremé residents have endured for decades. Brass bands routinely use the overpass as a destination for "second lines," and Sakakeeny offers an analysis of this phenomenon: "'Under the bridge' is what locals call the space below Interstate 10, and every jazz funeral and parade I attended in the downtown district of New Orleans wound its way there. Why? For one [thing], the concrete columns and 'bridge' overhead create an intimate space, enclosing parade participants, maximizing participation and a sense of collectivity. And the concrete creates spectacular acoustics, amplifying and multiplying the participatory sound, creating a sort of 'unplugged' feedback loop; acoustic, but also shockingly loud, and made louder by the musicians playing at peak volume to compete with the sound of cars and trucks whizzing by above."[29]

Philip Frazier, who co-founded Rebirth in 1983, commented on how bands conform to the acoustical and spatial possibilities of the streets: "When you get to a certain intersection or a certain street where there's an opening, if the street is really wide, you know that's more dancing room for everybody, you wanna keep everybody upbeat. When you get to a street where it's more closed, and the parade might slow down at a pace, you slow it down 'cause you know everybody's trying to get through that small street. . . . When you get under an overpass, 'cause of the acoustics, you know the band gonna be loud anyway, and the crowd knows that gonna be like some wild, rowdy stuff and you want to get everybody hyped."[30] Utilizing Steven Feld's concept of "soundscapes," Sakakeeny suggests that the manipulation of people, places, and sounds by brass bands to reclaim expropriated terrain, if only intermittently, encourages cultural unity and coherence within the Tremé community. Musicians and their followers convert what Henri Lefebvre calls "abstract space" (a matrix of bureaucratic regulations, such as parade permits, impeding cultural production) into "concrete space" (a site of living culture), which is another way of saying that, when brass bands and "second lines" are on the streets of New Orleans, they own that space.[31]

As it always has, New Orleans musical culture continues to grow by a process of eclectic experimentation in which tradition serves as raw material for reinvention, reasserting its relevance to a changing reality. Individuals operate within these parameters in different ways. TBC trumpeter Sean Roberts states: "Good music is good music. Sometimes we'll play, like, a hip-hop song with an uptempo beat, and some older guys, they will be like 'That's not traditional New Orleans music.' And sometimes we go, 'Kiss our ass, we know it's not. We know that.'" Young Fellaz trumpeter Ashton Hines continues: "As far as the fusion, you feel it. If I'm playing a traditional New Orleans song and a hip-hop song comes to my mind, I play it. It'll fit perfectly. I guess good music can do that."[32] Yet sometimes the message goes beyond strictly musical values, incorporating overt political statement. After Hurricane Katrina, one of the most vibrant underground music scenes in the city coalesced amidst the devastation of the Seventh Ward on A. P. Tureaud Boulevard, where brass bands and hip-hop performers gathered on weekends at a club euphemistically called "Duck Off." In performing their popular rendition of "Casanova" there, Rebirth routinely invited the crowd to substitute the lyrics "Fuck George Bush" in the chorus, referencing the chant made by angry crowds stranded at the Convention Center (and televised by CNN) immediately after Katrina.[33] Like the so-called "crack funerals" in Tremé, "Duck Off" was not intended to be a tourist destination, but it held meaning for the community as a site of memory and possibility. Yet, tourists will always have a place in the brass-band equation, not only because musicians need their dollars, but also because "second lines" are intended to be inclusive. Despite the predilection of some scholars to celebrate the "authentic" and repudiate "ersatz" versions of "second lines" for tourists, the musicians do not necessarily make such distinctions. In 2004 Benny Jones told me that Tremé Brass Band sees every "second line" for tourists as an opportunity to establish a human connection, to bring outsiders into the black culture of New Orleans so that they can understand how enjoyable (and necessary) such activities are to the people who live here. The musicians have found no contradiction, or shame, in embracing such multifaceted interpretations of tradition.

It is becoming apparent that the New Orleans brass-band community is more fluid and inclusive now than it has ever been. Many young musicians aspire to membership, which is why Rebirth snare drummer Derrick Tabb's Roots of Music program has a waiting list of more than 400, while currently accommodating 150 students. As in the past, the musicians will continue to work with the opportunities that the market affords them—often for love rather than money, although both are preferred. Thus, on October 23, 2010, there occurred an event promoted as the Red Bull Street Kings Brass Band Blowout, featuring battles under the Claiborne Overpass among the Free Agents, TBC, the Stooges, and the Soul Rebels.[34] There was a cash prize, plus three days at the Red Bull recording studio in Los Angeles.[35] We should not be deterred by the blatant commercialism apparent in this event; after all, capitalism is the American way, and these musicians were up for the challenge. The final round was a showdown between the Free Agents and the Stooges, with the crowd choosing the latter and the panel of judges concurring. Yet the Stooges' victory was controversial and sparked a vendetta that played out for weeks afterward.

In fact, the Stooges already enjoyed a reputation for inciting "humbugs" with other bands, notably several challenges to Rebirth that resulted in fights during the Big Nine annual "second line" in the Lower Ninth Ward in 2009. At the Red Bull event there were complaints that the Stooges had arrived wearing T-shirts bearing their name juxtaposed with the Red Bull company logo and had exceeded the permissible number of players on stage. Lack of clarity regarding the judging criteria also complicated matters, and when members of the Stooges appeared at a later Free Agents show carrying the oversized World Wrestling Federation–style belt that they had won at the Brass Band Blowout, it was interpreted as a provocation. Walter Ramsey (sousaphone and leader of the Stooges) recalled the incident: "I watched one guy grab the belt and stomp on it. He was so upset. I guess he thought we would be upset, too. The belt was just a gift, it's not that serious. When he saw it didn't bother us, he realized 'Wow I'm tripping. It's not that serious.' It kinda broke the ice to mend the situation."[36] At the Red Bull Blowout, host Glen David Andrews had concluded that "the true winners were the people of New Orleans... because we showed that violence has nothing to do with second lines," and while the aftermath of the event did lead to confrontation, resolution was achieved without bloodshed. The New Orleans brass-band community's faith in itself, based on a renewed sense of camaraderie, had proven to be stronger than its centrifugal tendencies, at least this time. As one observer commented, the Brass Band Blowout "was a beautiful event, and a perfect example of how an international corporation can come to New Orleans and stage an event that properly celebrates and promotes our local culture."[37]

What is clear in retrospect is that Red Bull felt it would benefit by an association with New Orleans brass-band culture, while also basking in the potential afterglow of corporate charity work. Considering this event in relation with the HBO series *Treme,* which is generating significant revenue for some New Orleans musicians, one might conclude that a positive pattern is emerging that may negate the dire prognosis offered by Sweet Home New Orleans in 2008. Hurricane Katrina focused international attention on the city's brass-band musicians and taught them that tradition cannot be taken for granted and must be open to change—like the population itself, it must be resilient, and its adherents must learn how to stand together. Today young New Orleans brass-band musicians continue to negotiate careers on their own terms aesthetically, if not always financially, and that bodes well for the future. Perhaps even more inspirational is the realization that scholars and musicians have found renewed common cause in fighting for a unique cultural patrimony in which variegation, excess, deviance, and singularity still have value. In a nation at risk of succumbing to an undifferentiated and mind-numbing mall culture, that achievement alone counts for a lot.

Notes

1. *Funk & Wagnalls Standard College Dictionary* (New York: Funk & Wagnalls, 1968), 229.

2. Standard works on the early development of New Orleans brass bands in jazz are William J. Schafer and Richard B. Allen, *Brass Bands and New Orleans Jazz* (Baton Rouge: Louisiana State University Press, 1977), and Richard H. Knowles, *Fallen Heroes: A History of New Orleans Brass Bands* (New Orleans: Jazzology Press, 1996). Knowles places the "pre-jazz period" from 1880 to 1900 and the "early jazz period" from 1900 to 1918. The "modern era," which Knowles dates from 1939, included "stylistic changes resulting from the development of the role of saxophones, as they replaced the alto and baritone horns, and freer rhapsodic solo trumpet style" and represents for him a decline in performance standards by the 1950s (7–8). His first chapter is titled "Brass Bands in the Community," which sets the tone for the work as a whole.

3. See William Carter, *Preservation Hall: Music from the Heart* (New York: W. W. Norton & Co., 1991); Mick Burns, *The Great Olympia Band* (New Orleans: Jazzology Press, 2001); and Bruce Boyd Raeburn, "The Atlantic New Orleans Jazz Sessions," liner notes booklet for the Mosaic boxed set MD4-179 (1998), esp. 4–7. Recordings by Atlantic Records of the Young Tuxedo Brass (1958) and the Eureka Brass Band (1962) helped to catalyze the entry of New Orleans brass bands into the tourist market. Over the course of the 1960s and 1970s, Preservation Hall developed the model for presenting "authentic" New Orleans music to tourists, based on national and international touring policies that expanded market exposure and attracted visitors to the city and the hall, where brass bands such as the Eureka and Olympia could be heard indoors—a departure from their community-based function as marching bands. See Bruce Boyd Raeburn, *New Orleans Style and the Writing of American Jazz History* (Ann Arbor: University of Michigan Press, 2009), 1, 255–58 for more information on Preservation Hall and "authenticity" issues.

4. For information on "second-line" street culture, see Helen Regis, "Second Lines, Minstrelsy, and the Contested Landscape of New Orleans Afro-Creole Festivals," *Cultural Anthropology* 14, no. 4 (November 1999): 472–504, and "Blackness and the Politics of Memory in the New Orleans Second Line," *American Ethnologist* 28, no. 4 (2001): 752–77.

5. For post-Katrina issues related to "second-line" festival traditions as catharsis, see Joel Dinerstein, "Second Lining Post-Katrina: Learning Community from the Prince of Wales Social Aid and Pleasure Club," *American Quarterly* 61, no. 3 (September 2009): 615–37, and Bruce Boyd Raeburn, "'They're Tryin' to Wash Us Away': New Orleans Musicians Surviving Katrina," *Journal of American History* 94 (December 2007): 212–19.

6. Rosalie Cohn, "Live for Today," *OffBeat*, October 2010, 28–34; quote on 29.

7. Music Rising was co-founded by producer Bob Ezrin, U2's The Edge, and Gibson Guitar CEO Henry Juszkiewicz in November 2005 to help Gulf Coast musicians affected by hurricanes Katrina and Rita. Phase One of its operation was a collaboration with the Grammy Foundation's Musicares program to provide instruments for New Orleans musicians who had lost them. Phase Two did the same for churches, schools, and Black Indian tribes. Phase Three is an award to Tulane University to develop an online curriculum focused on music of the Gulf Coast. For additional information, see www.musicrising.org/about.

8. During 2008, the nonprofit recovery organization Sweet Home New Orleans (SHNO) compiled a report on musician demographics, based on hard data compiled from questionnaires, which can be found at www.sweethomeneworleans.org/wp-contents/2008-csr-post-final.pdf. Basing its survey on client information (approximately 300–400 individuals), SHNO estimated that there were 4,500 musicians, Mardi Gras Indians, and social aid and pleasure club members living in New Orleans before Katrina impacted the city on August 29, 2005. See also *State of the New Orleans Music Community Report, 2010*, www.sweethomeneworleans.org, for subsequent iterations of this report.

9. See, for example, "New Orleans Panel Cites Jazz as Key to Rebirth," *San Francisco Chronicle*, January 17, 2006, D8. See also Howard Reich, "A Crisis of Culture in New Orleans: Battered by Katrina, the Cradle of America's Artistic Identity Might Never Recover Its Vitality," *Chicago Tribune*, July 2, 2006, 2, available at Lexis-Nexis Academic Universe, and Henry C. Lacey, "Other Opinions: Time to Face the Music—Blueprints, Visions for Rebuilding," *New Orleans Times-Picayune*, November 27, 2005, B15.

10. For information on the projections and perspectives offered by Mayor C. Ray Nagin's Bring New Orleans Back Committee in 2006, see Reich, "A Crisis of Culture."

11. For information on the range of the Grammy Foundation's Musicares programs, consult their website, www.grammy.org/musicares.html. Among the programs was an instrument-replacement initiative providing coupons for free instruments at Guitar Center, underwritten by NARAS. Music Rising used the Musicares database to identify recipients for its parallel instrument-restoration program.

12. For more information about the New Orleans Musicians Assistance Foundation of the New Orleans Musicians' Clinic, which served as a clearinghouse for a variety of charitable contributions, see www.neworleansmusiciansclinic.org.

13. For initial musician responses to Musicians Village protocols and subsequent developments, see Raeburn, "They're Tryin'," 815; Leslie Williams, "Habitat for Harmony: Musicians Village to Help Nurture Jazz Traditions," *Times-Picayune,* December 7, 2005, B1, B3; and Katy Reckdahl, "They Got It Bad," June 29, 2006, *OffBeat,* offbeat.com/artman/publish/article_1596.html.

14. The session, at which this author was present, was "Recovery in the New Orleans Community: A Report Card and Prospectus for the Future," Sixteenth Annual Cutting Edge Music Business Conference and Roots Music Gathering, Westin New Orleans Hotel Canal Place, New Orleans, August 15, 2008. For a review of the session, including details on the educational and instrument-supply activities of several of these organizations, see Dave Robinson, "The TJEN Corner: New Orleans Revisited," *Mississippi Rag* 35, no. 10 (October 2008), e-journal, www.mississippirag.com/ragonline_oct08/columns_tjen.html.

15. For a concise and insightful overview of the main lines of development of New Orleans hip-hop, see Ned Sublette, *The Year Before the Flood: A Story of New Orleans* (Chicago: Lawrence Hill Books, 2009), 185–227, which demonstrates how neighborhood and kinship affinities often connect New Orleans rappers and brass-band musicians as much as aesthetic predilections do.

16. As quoted in Jeremy Tolliver, "'No Place like Home': A Brass Band in Post-Katrina New Orleans," MA thesis, Bethel University, 2007, 39.

17. Lawrence Gushee, "The Nineteenth-Century Origins of Jazz," *Black Music Research Journal* 14, no. 1 (Spring 1994): 1–24, see esp. 11–12.

18. Tolliver, "'No Place like Home,'" 39–40.

19. Ibid., 28, 15.

20. Ibid., 20–21.

21. Michael G. White, "The New Orleans Brass Band: A Cultural Tradition," in *The Triumph of the Soul: Cultural and Psychological Aspects of African American Music,* ed. Ferdinand Jones and Arthur C. Jones (Westport, Conn.: Praeger, 2001): 69–96; quote on 89.

22. Keith Spera, "Facing the Music: Valuable Jazz Artifacts Drown in Floodwater," *Times-Picayune,* October 22, 2005: B1, B3. See also Michael G. White, "Reflections of an Authentic Jazz Life in Pre-Katrina New Orleans," *Journal of American History* 94, no. 3 (December 2007): 820–27. A short list of other notable musicians who experienced major losses due to Hurricane Katrina would include Dave Bartholomew, Fats Domino, Henry Butler, Pete Fountain, Bob French, Sybil Kein, Irma Thomas, and Al "Carnival Time" Johnson. In an effort to restore or replace damaged materials whenever possible, the Hogan Jazz Archive at Tulane University has provided oral history interviews, photographs, sheet music, and recorded sound artifacts to several of the affected musicians on this list, particularly Dr. Michael White and Sybil Kein, as well as to families of other jazz musicians who lost personal collections.

23. The matter of how musicians have been drawn into the brass-band tradition and educated deserves some attention. In the period 1885–1915 the itinerant New Orleans music "professor"

James B. Humphrey taught brass-band techniques to plantation workers throughout the hinterland along the Mississippi River, enabling country folk to migrate to the city and become professional musicians. Within the city, similar "hands on" pedagogical practices combining elements of amateur and conservatory traditions operated within families and neighborhoods. In most cases, only minimal rudimentary instruction was provided, and as often as not, participation in "second lines" from an early age becomes the actual prelude to musicianship. By the 1970s, Orleans parish elementary and secondary schools (and especially the New Orleans Center for Creative Arts, established in 1973) were routinely offering enough basic instruction to serve as a platform for aspiring musicians to try their luck on the streets of the French Quarter, which is how the Rebirth Brass Band began. Additionally, intervention by highly motivated individuals, such as Danny Barker, whose Fairview Baptist Christian Church Band experiments in the 1970s rekindled interest among youngsters in the tradition, generating brass-band projects ranging from the experimentation of the Dirty Dozen and Leroy Jones' Hurricane to more traditional units led by Michael White and Gregg Stafford, indicates the power of human agency in recruiting musicians and sustaining the tradition. For details, see Karl Koenig, "The Plantation Belt Brass Bands and Musicians, Part 1: Professor James B. Humphrey," *The Second Line* 33 (Fall 1981): 24–40; Al Kennedy, *Chord Changes on the Chalkboard: How Public School Teachers Shaped Jazz and the Music of New Orleans* (Lanham, Md.: Scarecrow Press, Inc., 2002); and Mick Burns, *Keeping the Beat on the Street: The New Orleans Brass Band Renaissance* (Baton Rouge: Louisiana State University Press, 2005). Most brass-band musicians working today have had some training in high school, and many continue to study music in college, but their finishing school remains professional networking in the streets. In the post-Katrina environment, where charter schools have essentially done away with music education, supplementary after-school programs with limited enrollments, such as Roots of Music, are the only current hope for aspiring youngsters seeking a more structured learning environment.

24. "Drummer's Funeral Underlines New Orleans Violence," *All Things Considered,* NPR, 6 January 2007. Hot 8 has not been the only brass band to suffer traumatic loss of its members. On May 9, 2010, TBC's twenty-two-year-old saxophonist Brandon Franklin was murdered in the Hollygrove neighborhood. See Cohn, "Live for Today," 32. For the correspondence of violence in New Orleans hip-hop, which parallels the experience of local brass-band musicians quite closely, largely because many of the same neighborhoods are involved, see Nik Cohn, *Triksta: Life and Death and New Orleans Rap* (New York: Alfred A. Knopf, 2005).

25. "New Orleans Journal: Café Culture," *New Yorker* blog, February 1, 2007, www.newyorker.com/online/blogs/neworleansjournal/2007/02/caf_culture.html.

26. *All Things Considered,* NPR, January 6, 2007.

27. Howard Reich, "A Culture's Sad Finale? Crisis of Culture in New Orleans," *Chicago Tribune,* February 4, 2007, www.chicagotribune.com/news/opinion/chi-0702040293feb04,0,665458.story.

28. Matt Sakakeeny, "Instruments of Power: New Orleans Brass Bands and the Politics of Performance," PhD diss., Columbia University, 2008.

29. Matthew Sakakeeny, "A Sound-Body Politic: Making Claims on Public Space through Sound," unpublished paper presented at the Society for Ethnomusicology conference, Columbus, Ohio, October 27, 2007, 3.

30. Sakakeeny, quoting from an interview with Philip Frazier, November 7, 2006, 10, unpublished paper.

31. Henri Lefebvre, *Rhythmnanalysis: Space, Time, and Everyday Life,* trans. Stuart Elden and Gerald Moore (New York: Continuum, 2004).

32. Cohn, "Live for Today," 30.

33. Anecdotal account based on personal conversations with Ariana Hall, founder of Cuba-Nola, who attended these events. For further exposition of hip-hop as a forum for post-Katrina dissent, see Zenia Kish, "'My FEMA People': Hip-Hop as Disaster Recovery in the Katrina Diaspora," *American Quarterly* 61, no. 3 (September 2009): 671–92, esp. references to 5th Ward Weebie's "Fuck Katrina (The Katrina Song)" (677–78) and Juvenile's "Get Your Hustle On" (686–87).

34. Red Bull's Brass Band Blowout project team, led by Field Marketing Manager Scott Lopker, spent over a year conducting research on the history of "cutting contests" in New Orleans jazz at the Hogan Jazz Archive at Tulane University in preparation for the event, including a final meeting with the author (acting in his capacity as curator of the archive) to discuss "authenticity issues" on August 4, 2009. The extent to which Red Bull was willing to go to create an event that would have credibility within the brass-band community was quite remarkable, and judging by local responses afterwards (see note 37, below), they succeeded.

35. Barbie Cure, "Street Fight," *OffBeat,* October 2010, 36–37.

36. Red Cotton, "Walter 'Whoadie' Ramsey, Leader of the Stooges Brass Band," blog of New Orleans, December 30, 2010, www.bestofneworleans.com/blogofneworleans/archives/2010/12/30/walter-whoadie-ramsey-leader-of-the-stooges-brass-band. The Stooges were formed in 1996 and, according to Ramsey, are "the most sought after, most controversial and most competitive band on the brass band scene," performing at the majority of "second-line" parades during 2010. "Just like their namesake [the Three Stooges], the band is known and adored for clowning during the shows, mugging at the audience, group dancing, inciting passionate city ward call and responses. . . . We have band members 17 to 33 years old. To have that group together, young and old, the fun is what makes it work." Ramsey studied at the New Orleans Center for the Creative Arts and is the grandson of a member of the Dirty Dozen Brass Band. His response to the hard feelings following the band's Red Bull victory was candid: "Those guys didn't understand this was a competition. I'm very competitive. I'm gonna out-think, out-play y'all 'cause that's what I do. I was kinda disappointed 'cause the Rebels are a great band. They shoulda made round two. Not to take away from the Free Agents, but the Free Agents don't know how to catch an audience. TBC, too. Sounding good is one thing, getting the people into it is another. . . . [I]t's about being an entertainer."

37. Ben Berman, "Let's Go Bluesin': Red Bull Street Kings Brass Band Blowout: Photo Slide Show," October 25, 2010, www.offbeat.com/2010/10/25/red-bull-street-kings-brass-band-blowout-photo-slideshow.

Memory Lives in New Orleans

The Process and Politics of Commemoration

SARA LE MENESTREL

When disaster strikes a community, the loss of family members and friends combines with a collective sense of loss and grief that afflicts all of the survivors through the destruction or damage brought to their homes, their possessions, and their memories. The ordeal of material and social recovery involves coping with the aftermath, which constitutes a part of the commemorative process. As the eminent anthropologist of disaster Anthony Oliver-Smith argues, the "ritual of mourning permits the bereaved to integrate the loss into their lives, to come to terms with it, and through the grieving process, resolve the conflicts inherent in loss between allegiance to the past and healthy reintegration into life."[1] The commemorative process provides assistance to people affected by loss in managing the transition from the grieving period toward the collective management of remembrance.[2] Sociologist Gaëlle Clavandier uses the notion of "collective death" as an object of study that traverses different kinds of deadly events such as wars, the Holocaust, or terrorist attacks, and requires a specific social treatment. Collective death does not depend on the scope of the disaster, but on the reactions and practices that it generates among a range of actors that includes authorities, media, experts, victims, and the population at large. According to Clavandier, "Each deed that involves several victims and that becomes an event requiring a specific ritualized treatment and a memory constitutes collective death."[3]

Whether transient or perennial, memorials act as vehicles for memory. They need not necessarily be cast in stone or concrete, but also include what has been called by Jack Santino "spontaneous" shrines, defined as nonofficial, "ephemeral" forms of mourning, emphasizing their transformative possibilities as acts of resistance.[4] Erika Lee Doss uses the notion of "temporary memorials": "The emotional life of public memorials is especially dependent on the fact that temporary memorials are created to be experienced: to be felt, not simply to be seen."[5] From this perspective, commemorative practices take a variety of forms. In New Orleans, official commemorations include the ritual ringing of the bells on August 29, at 9:38 a.m., marking the moment of the first levee break, the laying of wreaths in devastated neighborhoods, and the dedication of memorial monuments. Festive celebrations that are an essential aspect of the cultural landscape of the city, such as Mardi Gras, second lines, and even JazzFest, have assumed the character of commemorations, at least in the year following the disaster. Other large- as well as smaller-scale nonofficial practices function as tributes to the city and as assertions of a sense of belonging. These include bumper stickers, graffiti, bracelets, and tattoos, which stand as memorial devices that literally incorporate the pain.[6] Institutional or individual initiatives contribute to this commemorative process through the publication of countless essays and photo albums (by both New Orleanians and outsiders), documentaries, art exhibits, readings, and benefit and tribute music albums. The Internet has also become a rich vehicle for memory through the creation of public archives.[7] In fact, E. L. Quarantelli, a veteran of the field of disaster studies who created the pioneering Disaster Research Center at the University of Delaware in the 1960s, views these diverse commemorative practices as manifestations of a "popular culture of disaster," and he urges their further exploration by researchers in the field.[8] Together, these practices constitute a highly eclectic set of places, spaces, and moments where memory is convened.

Within the commemorative process, conferences have themselves provided a way for local academics, sometimes in collaboration with colleagues from out of state or abroad, to pay tribute to their native city and its population, to contribute to the healing process, and to show their support by applying their skills to a deeper understanding of the disaster.[9] Indeed, the workshop from which this book evolved was itself an integral part of this ongoing commemorative effort. However, the perspective of those among us who were residents of New Orleans at the time of the storm is necessarily marked by the scars left by this memory. The vivid experience of evacuating, returning, and living in a disaster zone—the emotions, feelings, images, and smells—lives in the memory of New Orleanians in ways that cannot be held at an objective distance.[10]

Anthropologists are routinely called upon to deal with the challenging combination of empathy and distance and of their status as both insider and outsider. One of our methods is to make the context of production of our work explicit through an introspective approach that necessarily frames and influences our research. Critical assessments of our own methods have become an integral part of our discipline, and in this spirit of reflexivity, we have a duty to take the commemorative character of this conference into account.

As for my own experience, I came to New Orleans in mid-October of 2005 to conduct fieldwork on the decision-making process of people affected by the hurricanes to return. After fifteen years of extensive fieldwork in Louisiana, my feelings of helplessness, loss, anger, and moral duty provoked this entry into the field of disaster research, a field that was totally new to me.[11] Enduring feelings of powerlessness and intrusion made fieldwork extremely uncomfortable.[12] I derived some comfort, however, from the people I interviewed acknowledging the benefit of telling their story, sometimes for the first time, and of having their words recorded by an outsider, a European, at a time when the city's very existence was being questioned by their fellow citizens. It is in this perspective, as a support for memory, that conducting fieldwork in a disaster zone became more tolerable both for me and for those I interviewed. Because my fieldwork ended after the first anniversary of Katrina, none of my numerous interviews up to that point addressed commemorative practices, and I was not able to study the use of memorials or the participants involved. Despite these gaps, however, my experience in the field during that first year and my participant-observation of the first anniversary provided a point of departure for our collective reflection.

For present purposes, commemorations are understood as multivocal, multifaceted practices, rituals, or ceremonies that mark the disaster. This definition excludes print and virtual memorials. I am interested only in objects and events that are meant to be seen, that occupy public space, and that take place locally, in New Orleans. Different stages of commemorative events, from the very aftermath of the disaster to its first and subsequent anniversaries, reveal practices that first emerged in 2006 and continue to be invested with remembrance. Within this perspective, I chose three forms of memorials: the house as the material expression of home, the Lower Nine Monument, and Katrina second lines. These practices use different objects as vehicles for grief, memory, and oblivion. Each dimension is given different priority depending on the time and place of the event and the actors involved. These different ways of framing memory express conflicted views of the disaster, of its implications, and of the ways to cope with it. Who is being memorialized and how? What is officially marginalized or forgotten, and what struggles are revealed? How is oblivion articulated, and according to what logic?

Katrina commemorations, rather than recalling what happened due to the hurricane, act mostly as responses to the multiple failures that inequitably affected the population and that exacerbated discrimination. This may explain the eclecticism of commemorative practices and the different meanings with which they are invested. Each practice seems to reflect the positioning of its organizers and acts as a space of contest.

Demolishing or Reconstructing?

"How do you commemorate the anniversary of something that is still happening? The devastation of our city is not just something that happened a year ago, it's something that was going on yesterday, continues today and will go on tomorrow," wrote New Orleans journalist Jordan Flaherty in August 2006.[13]

Through this statement that points out the endless character of Katrina, Flaherty illustrates the temporality of disaster as a process as opposed to an event. While up until the 1970s disaster studies focused on the ability to mobilize the affected population, American and French researchers have subsequently posited a "continuity principle," arguing that, in addition to disrupting social organization, disasters also reveal latent vulnerabilities and preexisting social divisions. According to Claude Gilbert, "the difference between the normal conditions and crisis situations is perhaps not as great as it would appear since there is less of a break than there is a continuity between the 'organized disorder' of the original condition and the 'disorganized disorder' of the aftermath."[14] The post-accident crisis, through its very characteristics, appears as "an 'excess' of the normal condition, this shift being sufficient to trigger significant disturbances."

From this perspective, the disaster appears to be more a cathartic factor than a force that destabilizes the social order. These analyses lead us to view the disaster itself as a social construction, a "revealing crisis."[15] This theoretical approach has decisive consequences for the research methodology to be adopted, because it leads us to focus less on studying an event than on revealing the conditions under which disasters are produced and on the "social creation of vulnerability."[16] According to a participant in an online forum, "Katrina, Katrina, Katrina . . . I live near the 17th Street Canal, very close to where the levee failed there on August 29, 2005. Katrina blew a piece of aluminum trim off of my house. It cost me exactly $200 to get it fixed. Because of that very minor damage, I had no lingering effects from Katrina. In fact, if the levees hadn't failed I probably would have to think about exactly what the name of that storm was back in late August 2005. . . . When the levee failed because the Corps of Engineers hadn't built them properly, now that was another story. I lost all of my worldly possessions, and now 5 years later I continue to work to get my life back to normal."[17]

Indeed, Katrina has been used as an umbrella event that covers much more than the hurricane itself, a use of which this resident disapproves because it places blame on natural causes instead of human responsibility. On the city's official web site dedicated to the fifth anniversary, there is no mention of the levee breaks, and Katrina is actually defined solely as a hurricane.[18] Other names given to the disaster by the population are far more explicit and have been taken up by the local media, including "the flood," "the federal storm," "the federal floods," "the levee breaks" or "failures." These labels point to the role of the government in this disaster before, during, and after the hurricane strike. Far beyond August 29 and the week that followed, the temporality of the disaster embraces an unfolding tragedy, particularly obvious and staggering one year after the event. As Rebecca Solnit puts it, we are faced with "a succession of disasters."[19] In fact, it took the mayor, city council, and civic leaders eighteen months to agree on a unified planning process with professional assistance, the Unified New Orleans Plan (UNOP), and on the preparation of a citywide infrastructure plan.[20]

A year after the disaster, about one-third of the population had returned to the city. Most schools and hospitals, particularly those serving the poor, remained closed and, overall, public services were operating at less than half capacity. There was no long-term planning for the city's reconstruction or for coastal restoration. Piles of debris continued to obstruct the sidewalks in the most devastated neighborhoods, and stacks of flooded-out cars were still under the interstate highway that traverses the city. Some corpses were even found six months after the waters receded. In other words, to the extent that all neighborhoods were taken into account, traces of the disaster were as visible on the date of the first commemoration as were signs of recovery.

The process of rebuilding was not only painfully slow, it began in fact with the bulldozing of twenty-five hundred of the most severely damaged homes. This demolition, which the Nagin administration undertook toward the end of December 2005 without contacting the homeowners, was experienced as a continuation of the destruction caused by the levee breaks in the most devastated neighborhoods, and it further deteriorated relations between the population and public institutions.[21] As a result, ACORN (Association of Community Organizations for Reform Now), one of the most active civic associations in the city in the defense of evacuees' "right to return" and assistance in home rebuilding, filed suit against the city.[22] "Memorials, not Demolitions" was their slogan, expressing the need for symbolic acknowledgment of grief long before the anniversary or the demolition of wrecked houses. On December 28, they won a settlement that required the city to notify homeowners of anticipated demolitions and to provide them with the opportunity to appeal. On April 6, the City Council passed an ordinance that allowed residents until the one-year anniversary of Katrina to clean and gut their homes before risking demolition. Once again, the use of the first anniversary as a deadline for demolition rather than a moment of remembrance reveals the gap between institutional and official priorities and those of the population.

Home as Memorial

"No Place like Home": A Sense of Place

At the 2006 JazzFest, although many New Orleans musicians remained in mostly involuntary exile, few missed the opportunity to pay tribute to their city. In fact, the atmosphere was filled with an exhilarating energy, and many bands made the crowd feel their loyalty to the city through their presence and through the unique echo of their songs, whether written for the occasion or invested with new meanings. When they sang "No Place like Home," the Soul Rebels—who were and still are established in Houston—started a call and response on "504," signing the numbers of the New Orleans area code with their hands as a rallying call that was particularly performative. Whether native or not, New Orleanians are deeply attached to their city.[23] All commemorative practices share an allegiance to the notion of home in all of its meanings, from the material house to the social network of interactions that it embodies. Countless commemorative publications, music albums, and memorabilia (T-shirts, bracelets, bumper stickers) have evoked "home" since Katrina and paid tribute to the city.

The notion of "home" overlaps with the notion of "community," also widely applied. Beyond the spatial unit, it refers to a network of relationships and mutual support that is assimilated to family ties. New Orleans is often referred to by its residents as "a small town," or by academics as "a collection of urban villages."[24] Indeed, one may be from New Orleans, but one was born or raised in a specific neighborhood, like Tremé, Gentilly, Uptown, Back o' Town, Mid-City, New Orleans East, the Ninth or Seventh Ward. The sense of belonging to a "close-knit community" and of always having the opportunity to run into friends is often mentioned by residents as an asset and, for nonnatives, as a substitute for deeply rooted family ties, in one of the most permanent populations in the country.[25]

Grieving Home

Three or four times . . . we were driving . . . I just pulled off the road; I just started weeping, just for no reason. There's a reason, I'm sure, but just uncontrollable. And Millie [his wife]'d asked, "What's wrong?" [He cries out]: My home!! Because it wasn't just a house, New Orleans was always my home. (Don Vappie)

Although a first visit four days after the storm reassured musician Don Vappie about his house in Covington—miraculously spared by the six trees that fell on his property—he recalls his sorrow during a subsequent visit in late September. For many, the confrontation with disaster after a more-or-less extended period of exile stands as the first occasion to experience grief and to release pain that had been repressed until that moment. Situations varied widely. Like Don, those who had been spared by the disaster were overwhelmed by emotions related to their dismantled home, in its broader meaning. Those who lost their houses had to confront various sights. Some came "home" to dwellings that were completely flattened or had floated or been driven some distance from their original emplacement; others were still standing but were totally decayed on the inside and infested with black mold. Wrecked houses appeared naked, as even their most intimate and trivial personal belongings were totally exposed to public view, dumped all over in a graphic violation of privacy.

The regulations that governed people's rights to return to their homes and the whole process of "look and leave," as well as the threat of having one's wrecked house gutted or even demolished, definitely exacerbated people's grief. In addition to their emotions, people had to deal with physical reactions: "The day me and my husband came back we came back to the house . . . [i]t was . . . I don't know what to say . . . my husband opened the door and we went in and he just started bleeding from the nose because of the odor. He couldn't take it. They had mold everywhere. The furniture was soaking wet. It was just horrible to walk into something knowing that it wasn't like that when you left it. We couldn't stay in there. We left right back out because we didn't come prepared for what we saw in there," recalled Gloria, a fifty-year-old member of the ACORN staff who lived in the Upper Ninth.

The pending decisions to let people return extended the grieving process from their first visit to cleaning and gutting. The limited time allowed for residents to see their houses and the required safety uniforms (gloves, masks, boots, as well as tetanus shot) contributed to a sense of dispossession, as did the large X-shaped markings left by authorities that branded every single building in the city, an extraordinarily powerful evocation of the disaster. What was left was no longer a house, it was *"something"*—like Gloria said above—an estranged place that had been intruded upon by natural elements and unknown human beings. Discovering one's ruined house was like experiencing death, and witnessing an absence or the phantom of something that was no longer present. In Spike Lee's documentary *When the Levees Broke,* Tanya, an active community organizer of the Lower Ninth Ward from ACORN, eloquently conveys her feelings: "It was like talking to a friend who had been disfigured. You know who you're talking to, you just don't recognize him."

Indeed, the flood caused a disruption of the boundary between built and natural environments. While debris was scattered in nature—such as belongings or even vehicles perched in trees—nature intruded on built-up spaces, such as mud and rafts of marsh grass and flotsam in yards and ruined houses. For some, the disruption was literally a matter of life and death in which the flood turned their homes, their living spaces, into coffins.

However, for most residents, after a first phase during which the experience of devastation actually enabled people to find closure, the healing process involved erasing all of the traces left by the disaster on their homes to collectively reengage in the present. Objects of everyday life became invested with particular meanings. In less affected neighborhoods, refrigerators and freezers stood silent sentry on sidewalks in front of most houses and apartments. From a distance, they lined up like the solid white crosses in a military cemetery. Wrapped with duct tape, sometimes even locked with a shackle or sprayed with images of skulls and messages like "Do Not Open," they carried the putrid contents in their forever-sealed guts. Acting as ephemeral memorials, refrigerators were used in a performative way, clearing out from the house a symbol of destruction and death. Many were also constructed as canvases for the expression of derision and protest, and in fact, whether spray-painted on refrigerators or on plywood panels or printed on T-shirts, humor has widely served as a coping mechanism while contributing to another kind of commemorative discourse.

In the most devastated neighborhoods, new, functioning houses have made the gaping holes where demolished houses have not been rebuilt seem even emptier, sometimes with orphan doorsteps or foundation pillars among the vegetation. The wrecked and abandoned houses, some unclaimed and overtaken by nature even now, five years later, seem all the sadder. Side by side with a neat, rebuilt house, the void immediately summons memory of loss and disaster, of people who did not return, or who died, while emphasizing the determination of the ones who returned. In this sense, and in the context of neighborhoods still struggling with recovery, the house itself—whether devastated, gutted, or restored—acts as a memorial, embracing both remembrance and oblivion. Indeed, the memorial erected in the Lower Ninth Ward used this symbol in support of the right to return.

The Lower Nine Monument

To reach the Lower Nine Monument from the city, one must cross the Claiborne Avenue Bridge. The site, laid out in a circle in the median, stands as a memorial to the people who lost their lives in the hurricane. On one side, ten blue poles show the flood levels in various parts of the city, while on the other side of the circle stands the frame of a newly constructed house (which could also be viewed as a gutted house) painted in red, on a gray slab with foundation and porch, symbolizing what has remained from the disaster. In the frame of the window is a sign: "I am coming home! I will rebuild! I am New Orleans!" Red chairs facing the house were meant by the sponsors to symbolize those who survived and returned, while a lone blue chair represents those who were lost to the flood. Further on, a paved walk leads to a marble gravestone in remembrance of "the victims and survivors of Hurricane Katrina and Rita," apparently seeking to give the memorial a broader, more regional reach.

Built over a period of just three weeks at the behest of Cynthia Ward Lewis, at the time a member of the Lower Ninth Ward Neighborhood Council and the City Council, the monument (whose materials, labor, and construction were donated) was designed by Boston-based architects Stull and Lee, Inc., "to honor the victims of Hurricane Katrina and to celebrate the indomitable spirit of the survivors determined to rebuild their community."[26]

Located in a median directly past the off-ramp of a high-speed road, with no crosswalks in sight, not far from the foot of a bridge that lacks pedestrian access, the memorial is designed to be seen from the window of a passing

Fig. 9.1. Lower Nine Monument. Photo by the author.

car or tour bus more than as a place that welcomes visitors on foot. In this sense, it emphasizes the isolation of the Lower Ninth Ward and its image of marginality, which dates from its first settlement in 1852. Built on marshland, the Lower Nine was further isolated from the city by the construction of the Industrial Canal, completed in 1923, which created a separation from the Upper Ninth.

The memorial's dedication illustrates the process of concealing not only the causes of the flood, but also the conflicts between the city and the state over recovery plans. On August 27, 2006, Cynthia Lewis led the ceremony to dedicate the memorial, surrounded by about fifty journalists, military, and firemen. A man read Matthew 14:22–33 and followed with a prayer: "God bless the breaches in our lives . . . the breaches of levees . . . of families . . . of homes. Put peace in our families, oh God, peace, be still." Although the mention of the levee failures was noteworthy, the focus was on the assertion of the right to return denied to the black population, and on the promise of rebuilding, against the original plans of the city's Bring New Orleans Back Commission. Governor Kathleen Blanco went on at great length, haranguing the crowd: "We're coming home.... No doubt that the Ninth Ward is coming home," she repeated. "This is an act of love and healing to remember, but we're gonna honor the past by building the future. It means building safer, smarter and stronger than before. Do I hear you?... so much of our culture was lost in the Lower Nine.... Do I hear you? The Lower Nine is coming back!"

Cynthia Lewis made similar promises: "Everyone who lived in the Nine who wants to come home can come home." In customary fashion dating from a series of blunt, awkward statements,[27] re-elected Mayor Ray Nagin attempted to throw in some humor, while identifying himself as an uncompromising black militant: "We plan to build N.O. bigger not smaller. I don't care who don't like it.... We gonna take any resources we can. Ladies and gentlemen, do not be discouraged, do not be bamboozled, do not be hoodwinked as Malcolm X used to say. This land is valuable. Build higher and better so that you can come back like nothing happened. Build your mansion, the first floors for cars, and the jacuzzi upstairs! [laughter]."[28]

Tacit enmities were visible, however: Blanco, who initiated the Louisiana Recovery Agency to establish a competing recovery plan, was announced as a "guest" in the printed schedule, and came second in Nagin's thanks after Lewis. In the crowd, several slogans were unambiguous about their claims and protests: "Stop ethnic cleansing" read a sign with blood dripping from the letters. Behind what appeared to be a united narrative, the official ceremony integrated the different dimensions that made it a multivocal and contested event. While official speeches were taking place, a black woman sitting in front of the monument was interviewed by Fox News. Outraged by the officials' rhetoric of political opportunism, she exclaimed: "Stop the talk, show me! I am very upset. Nothing is being done with the money. They're not rebuilding. It's not making sense. It's killing us, this mold and debris!... I'm licensed in environment. The water is not good. I'm sick of hearing the mayor's talk, councilmen.... If Katrina didn't kill me, they're doing a pretty good job!"

Indeed, the ceremony encapsulated the issues at stake at the time of the first anniversary: the active defense of the right to return by the population and grassroots civic organizations; the power struggles between the city and state administrations; the protest against the local government for its responsibility in the continuation of the disaster; the empowerment of homeowners; the insistence on the necessity of forgetting, and of looking toward the future. In fact, Gaëlle Clavandier characterizes official commemorative ceremonies as constructed on a curative logic of forgetting.[29] Forgetting the disaster is perceived as the only way to overcome the disorder caused by tragic

Fig. 9.2. Interviewed by Fox News during the official inauguration of the Lower Nine Monument on August 27, 2006, a woman expresses her outrage toward the city government. Photo by the author.

death and to return to "normalcy" ("like nothing happened" claimed Nagin). Memory is convened as a duty, but discourses on progress, improvement, and efficiency prevail.

The Imperative of Remembering Indiscriminate Loss of Life

The contestation of meaning and the rhetoric of forgetting called for other informal memorials to the victims of Katrina. On the ramp leading up to the St. Claude Avenue Bridge, the sole access for pedestrians to the rest of the city from the Lower Ninth Ward, people had spray-painted names (for example, "R.I.P. Lil Jimmy"), expressing their grief on a space that links the Lower Ninth to the city, highlighting its belonging and refuting its marginalization.[30] By 2010, however, the graffiti had been removed.[31]

During the year that followed Katrina, ceremonies were held to acknowledge loss, showing remembrance as an essential step in the recovery process.

The nascent Lower Ninth Ward Neighborhood Empowerment Network Association organized a public memorial service starting at the levee wall that included the reading of fifteen hundred names, and a walk from the levee to Flood Street past the Martin Luther King School.

The imperative to remember is best expressed by the Katrina Commemoration Foundation. Members explain why they organize a commemorative march every year by stressing the absolute necessity of "never forgetting" and "remembering" the lives that were lost, specifying the circumstances in which people died instead of speaking of the victims in broader terms: "We remember the suffering of those left in the sweltering heat without food and water on I-10, in the Convention Center and under the Causeway . . . we remember those killed by the police trying to save their families on the Danzinger Bridge, in Algiers and on Poland Avenue. . . . While we may want to forget Katrina, we cannot because we believe in a brighter hopeful positive future and we must learn from the past to build for the future."[32]

On August 29, 2006, the Katrina Commemoration Foundation organized a "healing ceremony" at the Lower Ninth Ward levee breach that would precede the "Katrina march/second line."[33] More than a hundred people walked to the levee, where two speakers declaimed the names of the deceased.[34] A group of women dressed entirely in white and wearing turbans and necklaces burned incense and chanted, accompanied by batá ritual drums. The leader of the ceremony was Olayela Daste, a minister of the Franklin Avenue Baptist Church.[35] She asked for a moment of silence before beginning to speak. She invoked various saints of the Yoruba (also called Orisha) religion, such as Ologun, deity of the depths, and Ogun, deity of war and technology, and also made reference to New Age energies ("the warrior") and Native American spirituality. After the ceremony, the Katrina march/second line paraded to Hunter's Field Park on the corner of North Claiborne and St. Bernard avenues in the Seventh Ward. The march culminated in an event sponsored by CNN, MTV, and other major companies, in which representatives of a dozen local and national civic, human rights, and African American organizations—among them the Hip Hop Caucus—were scheduled to speak before a concert given by fifteen hip-hop performers.[36]

By contrast with these eclectic commemorative ceremonies dedicated to the Lower Ninth victims that drew on a variety of registers (ranging from Christian services to ancestor worship, from spiritual remembrance to hip-hop tributes), the Lower Nine Monument did not seem to meet expectations of remembrance, and might not have been the object of a significant collec-

Fig. 9.3. Healing ceremony organized by the Katrina Commemoration Foundation, August 29, 2006. On the left, the leader of the ceremony, Olayela Daste, is chanting, accompanied by batá ritual drums. Photo by the author.

tive investment, although fieldwork on usage of the site would be necessary to support this assertion. Also, the fact that it is now referred to as "the Lower Nine Monument" seems to question its function as a memorial, a site of remembrance. Contested views over a proper large-scale official memorial are explicit in the efforts of the Katrina National Memorial Park, which claims to "build a real people's Memorial."[37] The project includes a wall of names in gold leaf of 4,500 to 5,000 victims, a number that contrasts with official figures of about 1,800. "New Orleans, the United States of America and the International community cannot afford to forget those loved ones we lost," proclaims the nonprofit association, positioning itself against the rhetoric of oblivion. An eternal flame set in a granite pool in the shape of a woman's womb further completes the language of loss and immortality. Furthermore, by associating the bond between New Orleans and its victims with the ties between a mother and her unborn child, the monument naturalizes the deep attachment of New Orleanians to their city.

While claiming an educational purpose through the use of cutting-edge green technology, this project is focused on the memory of lost lives and uses an emotional language of images and symbols through the monument and a mosaic of edited photographs of the disaster. It is noteworthy that in 2006 the Louisiana legislature approved a House bill establishing the Hurricane Katrina Memorial Commission, whose official goal was to make "recommendations for an appropriate memorial to commemorate those who lost their lives in Louisiana in Hurricane Katrina, as well as a site or sites for the memorial." However, no such memorial exists five years after the disaster. Rather, a variety of permanent official memorials have been erected in different towns and locations, among which are a cross in Shell Beach, a sculpture in Slidell, a tombstone at the Convention Center, and mausoleums at the Charity Hospital Cemetery.

Located near City Park between Saint Patrick Cemetery and the New Orleans Archdiocese Cemetery, the Katrina Memorial of the Charity Hospital Cemetery is little known within the neighborhood, which is the site of multiple cemeteries. This memorial is laid out in the shape of a hurricane with the idea of creating "a healing shape out of the form that brought us such pain," according to the originator of the concept, Dr. Jeffrey Rouse. Rouse is a psychiatrist employed by the Orleans Parish Coroner's Office, which is responsible for unclaimed bodies. The memorial is also shaped like a labyrinth, so that a visitor may wander and meditate among the unmarked mausoleums.[38] This opportunity for meditation is supported by benches that face the reflective surfaces. The combined reflection on these polished walls of a person seated on the bench and the planted beds highlights the contrast between the mobility of the leaves and branches and the immobility of the mausoleums. This sight makes people feel the interaction between life and death and the reintegration of the dead into the living. The location on the grounds of Charity Hospital, which before Katrina provided health care to low-income residents, restores dignity to the indigent and nameless deceased. The anonymous memorialization of unclaimed victims also broadens the spectrum of remembrance, "like a Tomb of the Unknown Soldier, emphasizing indiscriminate loss of life," as *New Orleans Times-Picayune* journalist Doug McCash expressed it.[39] Walking around these massive mausoleums is in fact a humbling experience.

Fig. 9.4. Mausoleum reflection of a bench, labyrinth, and plantings at the Katrina Memorial, Charity Hospital Cemetery. Photo by the author.

The site also clearly displays contested views of the disaster. As soon as one comes through the entrance gate, one sees a series of tombstones with texts signed by various officials. Nagin's tribute refers to "those who perished as a result of hurricane Katrina" on an initial series of tombstones that includes the memorial's many supporters. A large black, central tombstone that matches the mausoleums and stands in their midst is signed by the coroner's office, with a wording different from Nagin's and giving details about the "failures of the levee system" and the "desperate circumstances" and "harrowing days" faced by the population.

Living Memorials

The Stigmatization of Public Housing

It is no coincidence that the Lower Nine Monument anchors the memory of victims and the will to return into a neighborhood of homeowners. It is a neighborhood where 59 percent of households were occupied by owners, a rate that is higher than the city itself (46.5 percent) and among the highest ownership rates in black-populated neighborhoods in New Orleans.[40] Contrary to the image of irresponsible renters that is associated with the black population, the Lower Nine's population makes a point of emphasizing homeownership as a way to legitimize the right to return and the reconstruction of their neighborhood. The slogan "I'm from the 9 and you ain't takin' mine," used by ACORN on T-shirts and worn by many residents, captures the significance of the notion of homeownership. At the heart of the concepts of individual achievement and self-help, it is also inseparable from the definition of American citizenship.[41] The Lower Ninth became a symbol of social promotion and empowerment for African Americans in New Orleans, used as a strategic tool for survival and for asserting their rights as full-fledged citizens. This is one of the reasons why the Lower Ninth came to symbolize injustice and discrimination against the black population in the national and international press. The denial of homeowners' rights to return and rebuild triggered unanimous protest. Several nonprofit associations exclusively dedicated to the rebuilding of the Lower Ninth have been funded since Katrina, adding to those that already existed.[42] Homeowners also became the focus of rebuilding projects that have received international media attention, such as Habitat for Humanity's Musicians' Village and Make it Right, which has publicized the number of residents who "have achieved their dreams of homeownership" through these programs.[43]

Allusions to homeownership were frequent during Katrina's first anniversary. At the ceremony dedicating the Lower Nine Monument, Cynthia Lewis and Ray Nagin explicitly addressed homeowners. The slogan "I'm coming home! I will rebuild! I am N.O.!" was incorporated into the memorial and posted on individual houses, positioning homeowners as the fabric of the city and as the driving force of reconstruction and of the fight over the right to return. At Hands around the Dome, a ceremony organized by the African American Leadership Project and sponsored by the NAACP, Nagin repeatedly contrasted homeowners with criminals. City Council President Oliver Thomas's comment that public housing should be for people who work instead of "soap opera watchers"—he later apologized for the remark—completed the stigmatization of renters as second-class citizens.[44]

The closing of four extensive public housing complexes in New Orleans in 2007 made the need for low-income housing even more critical and revealed the blatant racial discrimination behind housing decisions and policies. St. Bernard Parish has since Katrina fought desperately to prevent the creation of housing for low-income families and an influx of renters, at one point even approving a law, later repealed, that prohibited homeowners—who are overwhelmingly white—from renting to anyone other than a blood relative.[45] Violence and dependency make up the distorted public image of public housing and contribute to the stereotyping of renters as opposed to homeowners.[46]

Second Lines and Jazz Funerals as Living Memorials

On January 15, 2006, a coalition of more than thirty clubs united to form the Social Aid and Pleasure Club Coalition Task Force. After four months of forced exile, thousands of displaced New Orleanians came from Baton Rouge, Houston, Dallas, and Atlanta (and some from as far away as San Francisco, New York, and Portland) to participate in a second-line parade and to check on their home, neighborhood, family, and friends. For some, it was their first return to the city since the storm. The "All-Star Second Line" drew over eight thousand people, who wore black T-shirts that read "Renew New Orleans." The parade functioned as a reunion and as a way to reclaim the streets.[47]

Second-line parades are emblematic of New Orleans culture. They are organized and paid for by social aid and pleasure clubs, the descendants of nineteenth-century benevolent and mutual aid societies that provided health, unemployment, and burial insurance for their members. They developed the tradition of jazz funerals, wherein a brass band plays dirges on the way to the cemetery, and then picks up the tempo for fast tunes. Today, most clubs hold an annual second-line parade in which the club members and the brass band constitute the first line and everyone else is a second-liner. While the All-Star Second Line was attended by a socially eclectic crowd and was invested with a special meaning since it was the first since Katrina, second lines usually gather a more homogeneous population: "The majority of participants in the second-line tradition are not owners of homes, real estate, or large, public businesses. Yet through the transformative experience of the parade, they become owners of the streets, at least for the duration of their performance," notes Helen Regis.[48] Whereas homeowners have been the focus of most commemorative ceremonies, second lines provided non-homeowners an opportunity to enact collective property by claiming popular ownership of public space. The unified second line that was held after Katrina, as well as other second lines and jazz funerals dedicated to the disaster, have all functioned as commemorative practices. Returning to the normalcy of second lines as central cultural events within New Orleans's African American working class took on a special meaning at a time when the fencing off of major public housing complexes was unfolding. Through second lines, social aid and pleasure clubs reasserted their civic engagement and crucial leadership functions for the most disadvantaged African Americans.[49]

Second-line parades held after the disaster also evoked the maintenance of a sense of continuity and normality through a festive calendar that contributes to the social fabric of the city. The unseen aspect of the disaster was thus transformed into the ordinary of African American expressive culture, apparently serving as a strategy for retaking control of commemorative practices. In this process of reappropriating devastated neighborhoods and reclaiming streets kept in desolation, in this struggle against dispossession, second-line traditions stood in the front line.

Tuesday, August 29, 2006. A Katrina jazz funeral is organized by the Black Men of Labor and the Popular Ladies social aid and pleasure clubs. The Tremé Brass Band provides the music, followed by first responders and a horse-drawn hearse. The route, from the Convention Center to the Superdome, situates the event not as a remembrance of the storm, but rather as a remembrance of the thousands of people who were left in hellish conditions at these two sites because the government failed to provide basic assistance. Forty-five minutes passed before the music actually started, giving time for a huge, eclectic crowd to gather, including numerous journalists. A flyer was passed around with a long text condemning violence as entirely opposed to the club's ethic.[50] As the parade proceeded, the music became more and more upbeat, and jubilant participants started to sing along furiously. An overwhelming need to take part in this parade could be sensed among the marchers, and despite the burning sun, people did not miss a beat, dancing, hammering the streets, and raising their fists while singing call and response. Everybody was either wearing Katrina T-shirts calling for action and expressing blame of various kinds (from "Fuck Bush" to "The job's not done yet") or displaying tattoos dedicated to Nola.

While in the aftermath of Katrina it was not unusual to see ruined houses with spray-painted messages insulting Katrina, symbolic jazz funerals dedicated to the disaster have had a soothing function, helping to put an end to anger, anxiety, and uncertainty caused by death. The horizontality of the casket plays on the idea of resting through the prone posture of a person asleep.[51] As in any other funeral ceremony, jazz funerals mark a ritualized separation between life and death, allowing the grieving process to start and, eventually, opening the way toward a reincorporation of the dead into life. However, the message here was not one of oblivion. As opposed to official slogans from state authorities that pressed people to focus on reconstruction (Re-cover, Rebuild, Rebirth), the Black Men of Labor had their own version of the ubiquitous "triple R." "Remembrance, Renewal, Rebirth" said their banner, acknowledging the necessity to grieve, and making memory a condition of the reconstruction process. In fact, for the jazz funeral sponsored by French Quarter Hotel Monteleone on the occasion of the fifth anniversary, a sign on the casket read: "It's time to put the old girl to rest," as if to say that, five years later, the separation phase was still not complete. Indeed, the recurrence of disaster in the form of the BP oil spill can be symbolically interpreted as the failed neutralization of the death represented by Katrina. Katrina needed to be buried in 2010 once and for all, as though the specters of the deceased were still wandering about and might contaminate the living through epi-demic death.[52] In fact, the oil spill was symbolically incorporated into the jazz funeral through a BP tool box that was thrown into the coffin.[53]

The jazz funeral's metaphoric progression from death and mourning to celebration has naturally been applied to the recovery of the city. Other jazz funerals to honor the deceased—both literally and symbolically through the death of the city—were organized after Katrina. Helen Regis has shown how the names, images, and memories of the deceased have been integrated into some social and pleasure clubs' anniversary parades, in particular through the wearing of memorial T-shirts.[54] Indeed, at the 2006 Hands around the Dome ceremony, several children wore memorial T-shirts designed with a color photograph and the following caption, indicating the nickname and birth and death dates of the deceased:

MEMORY LIVES IN NEW ORLEANS
AFTERMATH OF KATRINA

Darryl G. Milton aka "Spook"
R.I.P.
Sunrise 4-6-1977 Sunset 8-29-2005

By invoking the living dimension of memory, this T-shirt emphasized the significance of "wearing the pain," and represented a direct reference to second-line practice as a living memorial. "Through the practice of designing and wearing memorial shirts," argues Regis, "second liners' very bodies be-come living memorials to the dead, as the images of those who have gone be-fore are paraded through the streets of the city, traversing from five to seven

Fig. 9.5. A group of children wearing a memorial T-shirt at the Hands around the Dome ceremony, organized by the African American Leadership Project and sponsored by the NAACP, August 27, 2006. Photo by the author.

miles of cityscapes in embodied grief. Through the wearing of memorial shirts, the entire parade is transformed into a memorial. . . . Unlike the cool and stationary cement, glass, and stone memorials of endowed buildings, the tactical memorials of New Orleans second liners are moving monuments made of flesh and blood."[55] Through living memorials, the dead and the living are inextricably tied together. Jazz funerals stand as a performative practice through which the dead materialize—whether through a casket, a T-shirt, or a photograph—and are reintegrated into the living, contributing to the completion of the grieving process. During this reunion, "the funeral is concluded on the celebration of a new order, where death gains a positive status, regenerating the community and the bereaved who become an integral part of the living again."[56] Second lines, as an already existing commemorative practice and central cultural event among African Americans, act as a source of empowerment for those who were dispossessed and as a continuously renewed commemoration of their role in the social fabric of New Orleans.

Conclusion

Based on a logic of oblivion, "commemorative memory," as borne by public officials, needs to carry a positive message for the present, argues Gaëlle Clavandier.[57] While its project is to create the conditions for a return to normalcy, what she calls "event memory," based on shared experience and emotions, has no clear project per se other than to continue living. These two types of memory, she argues, are both necessary and coexistent.

Whereas public officials' discourses are constructed on a logic of oblivion, the eclectic public commemorative practices in New Orleans share a language of remembrance. They express the imperatives of reparation and of healing, of remembering all New Orleanians regardless of their social status. They also assert the reclaiming of collective ownership and the restoration of dignity both by and for those who are indigent and dispossessed. The life of memory expressed by these commemorative practices gives voice to the reintegration of the dead into the living, a condition for the reconstruction process, in every sense of the word.

Notes

1. Anthony Oliver-Smith, *The Martyred City: Death and Rebirth in the Andes* (Albuquerque: University of New Mexico Press, 1986), 190.

2. Ibid.; Christian Delécraz, Laurie Durussel, eds., *Scénario Catastrophe* (Geneva: Musée d'Ethnographie de Genève, In-Folio, 2007).

3. Gaëlle Clavandier, *La Mort Collective: Pour une Sociologie des Catastrophes* (Paris: CNRS Editions, 2004), 186.

4. Jack Santino, *Spontaneous Shrines and the Public Memorialization of Death* (New York: Palgrave Macmillan, 2006).

5. Erika Lee Doss, *The Emotional Life of Contemporary Public Memorials: Towards a Theory of Temporary Memorials* (Amsterdam: Amsterdam University Press, 2008).

6. See Glenn W. Gentry and Derek H. Alderman, "Trauma Written in the Flesh: Tattoos as Memorials and Stories," in *Narrating the Storm: Sociological Stories of Hurricane Katrina,* ed. Dani-

elle Antoinette Hidalgo and Kristen Barber, 184–97 (Newcastle upon Tyne, U.K.: Cambridge Scholars Publishing, 2007).

7. Such web sites dedicated to oral history were created for other disasters such as the 2004 Asian tsunami, www.tsunamistories.net/.

8. E. L. Quarantelli and Ian Robert Davis, "The Disaster Research Center Working Paper # 92: An Exploratory Research Agenda For Studying The Popular Culture Of Disasters (PCD): Its Characteristics, Conditions, and Consequences" (Newark, Del.: Disaster Research Center, 2010).

9. This was the case of the 2006 spring meeting of the Society for Anthropology in North America, held in New Orleans and dedicated to "Unnatural Disasters." At Tulane, the 2009 international conference "Moving On: Trauma and Memory in History" was initiated as an "urgency" in response to "a patient still in intensive care," as stated by the conference program.

10. Oliver-Smith, *The Martyred City,* 26–30

11. Jacques Henry and Sara Le Menestrel, *Working the Field: Accounts from French Louisiana* (Jackson: University Press of Mississippi, 2009), xxx et seq.

12. Ibid., xxxii.

13. Jordan Flaherty, "New Orleans, One Year Later," *AlterNet,* August 29, 2006, www.alternet.org/wiretap/40976/.

14. Claude Gilbert, *Le Pouvoir en Situation Extrême: Catastrophes et Politique* (Paris: L'Harmattan, 1992), 247.

15. Anthony Oliver-Smith, "Anthropological Research on Hazards and Disasters," *Annual Review of Anthropology* 25 (1996): 304.

16. Ibid., 314; Sandrine Revet, "De la Vulnérabilité aux Vulnérables: Approche Critique d'une Notion Performative," in *Vulnérabilités Sociales, Risque et Environnement: Comprendre et Évaluer,* ed. S. Becerra and A. Peletier (Paris: L'Harmattan, 2009): 89–99.

17. *Nola.com,* forum, "How Did Hurricane Katrina Change You? 5 Years Later," tuffcookie, August 17, 2010.

18. See www.katrinafive.com/.

19. Rebecca Solnit, *A Paradise Built in Hell: The Extraordinary Communities That Arise in Disaster* (New York: Viking, 2009), 234.

20. Stephen P. Leatherman, Shirley Laska, Robert W. Kates, and Craig E. Colton, "Reconstruction of New Orleans after Hurricane Katrina: A Research Perspective," *Proceedings of the National Academy of Sciences* 103, no. 40 (2006): 14653–60.

21. In old Yungay after the 1976 earthquake and avalanche, Oliver-Smith (*The Martyred City,* 109) observed bulldozing of shacks reconstructed on the side of the road that caused a similar distrust.

22. Created in Little Rock, Arkansas, in 1970, ACORN has its headquarters in New Orleans, where the association included 9,000 members before Katrina. The organization has more than 300,000 members and chapters in one hundred cities across the country.

23. Sara Le Menestrel and Jacques Henry, "'Sing Us Back Home': Music, Place, and the Production of Locality in Post-Katrina New Orleans," *Popular Music & Society* 33, no. 2 (2010): 179–202.

24. Jason Berry, Jonathan Foose, and Tad Jones, *Up from the Cradle: New Orleans Music since World War II* (Athens: University of Georgia Press, 1986), 3.

25. Over 77 percent of residents were born in Louisiana and have lived most of their lives there.

26. See www.stullandlee.com/pdf/Stull and Lee Inc.

27. One famous example that aroused virulent reactions in New Orleans and the rest of the nation is an excerpt of Nagin's speech at City Hall to commemorate Martin Luther King Jr.: "It's time for us to rebuild a New Orleans, the one that should be a chocolate New Orleans. And I don't care what people are saying Uptown or wherever they are. This city will be chocolate at the end of the day. This city will be a majority African-American city. It's the way God wants it to be."

28. Nagin also endorsed this role the previous day, at the Hands around the Dome ceremony, where he was dressed in a dashiki.

29. Clavandier, *La Mort Collective,* 128.

30. Brian Rosa, "Tours and Detours: Walking the 9th Ward," *Triple Canopy* 3, NOLA special issue, canopycanopycanopy.com/3.

31. Other graffiti on the levee wall, where it had been breached, was covered with gray paint by Fred Radtke, known as "the Gray Ghost." Radtke has fought a battle against graffiti in the Crescent City since 1997, covering all graffiti and street art in the city, whether on public or private property.

32. See www.akinle.com/katrinacommemoration.

33. The 2006 and 2007 marches were hosted by the People's Hurricane Relief Fund (PHRF). In 2008, the New Orleans Katrina Commemoration Foundation was established to continue the march/second line.

34. In 2010, people were holding a long black banner with the list of deceased.

35. The name "Olayela" indicates that she was initiated into Ola, deity of joy.

36. Founded in 2004, the Hip Hop Caucus sees its mission as "to foster civic engagement among young people of color on issues of social and economic justice, human rights, the environment, and international peace." A brief video clip of the 2010 ceremony can be seen at www.hiphopcaucus.org/katrina.

37. See www.katrinanationalmemorialpark.com/photos.html.

38. "Dispatch: Reflecting on Anonymity at the New Orleans Katrina Memorial," comments by Adri Wong, September 21, 2010, *The Hydra,* www.thehydramag.com/2010/09/21/.

39. Doug McCash, "New Orleans Katrina Memorial Is Almost Perfect," September 14, 2007, Nola.com, blog.nola.com/dougmaccash/2007/09/new_orleans_katrina_memorial.html (accessed October 15, 2010).

40. Higher rates of homeownership are found in Pontchartrain Park, Gentilly Woods, and Gentilly Terrace.

41. Thomas J. Sugrue, *The Origins of the Urban Crisis: Race and Inequality in Postwar Detroit* (1996; Princeton, N.J.: Princeton University Press, 2005).

42. Among them are the Lower Ninth Ward Homeowners Association, the Lower Ninth Ward Neighborhood Empowerment Network Association, and Lowernine.com.

43. Such is the case of Make it Right, with the building of 150 houses in the Lower Ninth.

44. In fact, data from the 2000 census show that the majority of public-housing residents worked. Tenants earning a wage or salary ranged from 37.9 percent in Iberville to 69 percent in the Florida development. Employment in St. Bernard and Cooper were both 60 percent. See Bill Sasser, "'This Is Our Home': Tenants of Public Housing across New Orleans Feel Frozen Out and Unwelcome—at the Hands of Their Landlord, HANO," *Gambit,* April 11, 2010, www.bestofneworleans.com/gambit/this-is-our-home/Content?oid=1245574.

45. Campbell Robertson, "Housing Battle Reveals Post-Katrina Tensions," New York *Times,* October 4, 2009.

46. A notable exception to this discourse was the Care for Community Campaign on the fifth anniversary of Katrina. Chaired by former mayor Marc Morial, it claimed to "speak for renters" (www.5thkatrinaanniversary.info/).

47. Rachel Breulin and Helen Regis, "Putting the Ninth Ward on the Map: Race, Place, and Transformation in Desire, New Orleans," *American Anthropologist* 108, no. 4 (2006): 744–64.

48. Breulin and Regis, "Putting the Ninth Ward on the Map," 756.

49. Frederick Weil, "The Rise of Community Engagement after Katrina," *The New Orleans Index at Five* (New Orleans Community Data Center & Brookings Institution, 2010).

50. This claim of nonviolence refers to the controversy over the increase in permit fees for second-line parades. The New Orleans Police Department raised the fees by as much as 530 percent, arguing that more security was needed following two shooting incidents during second lines in January 2006. The clubs filed lawsuits against the city and won the case.

51. Clavandier, *La Mort Collective,* 135.

52. In St. Bernard Parish, a "Hurricane Katrina Casket" sponsored by the St. Bernard Memorial Funeral Home was made available to place notes about memories into the casket.

53. From this perspective, Gaëlle Clavandier notes the social and psychological dangers of floating corpses. See Clavandier, *Sociologie de la Mort* (Paris: Armand Collin, 2009), 60.

54. Helen Regis, "Blackness and the Politics of Memory in New Orleans Second Lines," *American Ethnologist* 28, no. 4 (2001): 752–77.

55. Ibid., 765.

56. Clavandier, *Sociologie de la Mort,* 86.

57. Ibid., 121.

Why Mardi Gras Matters

RANDY J. SPARKS

Hurricane Katrina was a monster storm. Its devastating winds of 115 to 130 miles an hour extended over 100 miles from its center. Initially it appeared that the city had avoided a direct hit, and the New Orleanians who had evacuated and those who were left behind breathed a sigh of relief. Then the city's man-made levees began to breach, and within eighteen hours almost 80 percent of the city was flooded under six to twenty feet of water. The floods, a result not of the storm but of poor construction and maintenance, destroyed hundreds of thousands of homes and left over 1,000 residents of the city dead, over 70 percent of them over the age of sixty. Over 1 million people evacuated, and over 770,000 were displaced, the largest number of Americans forced out of their homes and communities since the Dust Bowl migrations of the 1930s. The images of the tragedy that unfolded—people trapped and dying on rooftops; evacuation centers like the Superdome and the Convention Center spiraling into chaos; scenes of looting, violence, and disorder—were seared into the national consciousness and cast in high relief the failure of the government at every level to respond to the unfolding tragedy.

Six months later, the city still looked like a war zone. Well over half its residents had not returned, and the Federal Emergency Relief Agency (FEMA), the arm of the federal government charged with leading the recovery effort, had become a byword for inefficiency and bumbling incompetence. It was increasingly clear that the city's very survival was at stake, and in that context a debate began in New Orleans and around the country: should the city hold Mardi Gras in 2006? The festival would begin only six months after the storm, with the city still on its knees. A New Orleans magazine summed up the debate: "Elsewhere, many are shaking their heads in befuddlement, wondering how we can even think of throwing a party at such a time. Our city was ravaged by Hurricane Katrina; . . . vast portions of New Orleans won't be rebuilt; and worst of all, many in other parts of the country still think we brought this on ourselves or that we're not worth saving. Or both. In the face of such immeasurable loss, suffering and uncertainty, and with all the work ahead, how can we take time out for Mardi Gras?"[1]

In the wake of Katrina, New Orleans was a postapocalyptic society, having experienced devastation so severe, so cataclysmic, that it was pushed to the very brink of destruction. New Orleans was not destroyed; it survived, but in the midst of terrible loss and damage, there were real questions about how it would survive and what form that survival might take. In postapocalyptic societies the pre-disaster state is either forgotten or mythologized, the infrastructure is seriously compromised, and the people suffer mental and physical effects. But societies that emerge from an apocalyptic event can only do so by confronting the problems and finding some way to overcome them or live through them. In all such cases, a society's cultural heritage also faces destruction, and often the preservation of that heritage is not something that figures prominently in rebuilding efforts, particularly in the view of outsiders and government relief agencies. In this context it is crucial to appreciate that cities perform a number of functions that cover the whole range of human experience in an urban setting. Among those functions is *performance,* "actions undertaken for purposes of role clarification, identity confirmation, novelty, and experimentation rather than for purposes of direct adaptation and survival." Disaster recovery efforts have focused almost exclusively on the material and economic functions of cities and have shown little sensitivity to the other major aspects of urban life. Scholars who study the impact of disasters, particularly cultural anthropologists, have begun to understand the loss that disasters inflict on communities as a whole, and they have emphasized that it is an essential part of recovery for individuals and communities to grieve for lost homes, culturally significant places and buildings, gathering places, and the whole range of physical features that make a community. They have also found that in the wake of disasters survivors often hold onto their cultural traditions, and search for meaningful ways to commemorate suffering, loss, grief, and tragedy.[2]

The debate surrounding Mardi Gras shows how the people of New Orleans found the strength to return and rebuild their city and its rich cultural traditions. As New Orleans struggled to recover from Katrina, it often faced criticism from outsiders, and in large part that criticism was based on the city's reputation for immorality and corruption, a view of the city that had been shaped in large part by Mardi Gras itself.

Tradition holds that Mardi Gras came to New Orleans with the early French settlers, but hard evidence of the celebration of Carnival is scarce for the early period. Certainly Carnival has deep roots in Catholic Europe, where it began as early as the fifteenth century as a period of feasting and merriment beginning on Twelfth Night or the Epiphany and ending at midnight on Shrove Tuesday, the beginning of Lent. Mardi Gras, or Fat Tuesday, marked the last opportunity to feast before Lent. The French celebrated Carnival from the Middle Ages with feasting, dancing, drinking, sports, and with rituals mocking the social order. A man would dress as a "king of fools" or "lord of misrule," a tradition most popular in France, where men and women would parade as priests and nuns and obscenely parody the church, and cross-dressing was common. As Barbara Ehrenreich observed, "Whatever social category you had been boxed into—male or female, rich or poor—carnival was a chance to escape from it." Gradually, these celebrations became increasingly secularized, which gave the people more control over them. Goethe captured this essence of European Carnival when he described it "as a festival that really is not given to the people, but one the people create for themselves." Going back to the ancient Greeks, observers of festive public celebrations have noted that such ecstatic performances have a curative effect on participants. Ehrenreich suggests that communal celebrations reconnect individuals to their community, and that shared joy helped bring people out of themselves and created a sense of oneness with other celebrants. In early modern Europe, authorities tried to rein in Carnival and other rowdy celebrations, and the Protestant Reformation swept away many of the festivals associated with the Catholic Church calendar, but not in Catholic countries like France where Carnival continued to be celebrated.[3]

Supposedly, the French introduced masked balls to celebrate the season in colonial New Orleans, and certainly the earliest evidence points to their existence as the central feature of Carnival in the French period. The earliest documented evidence for Carnival comes from 1730. During the Spanish period in 1781 the City Council or Cabildo heard complaints that blacks were masking and slipping into Carnival balls. The Cabildo made it illegal for blacks to mask, to attend balls, or to wear feathers during Carnival. It is clear that, by the time of the U.S. takeover of Louisiana, Carnival was well entrenched in the Creole, or native-born, French-speaking community. American authorities banned weapons from public balls, required policemen to be present, and banned masking. Supposedly, a group of young Creole gentlemen returned to New Orleans in 1827 from their studies in Paris, and brought with them some Carnival customs, including public parading in costume. In 1827 the city council lifted its ban on masking between January 1 through Mardi Gras Day, with the lone dissenting vote cast by an American-born councilman. Street parades became a common feature of Carnival. James Creecy, who visited the city in 1835, described the scene: "Men and boys, women and girls, bond and free, white and black, yellow and brown, exert themselves to invent and appear in grotesque, quizzical, diabolical, horrible, humorous, strange masks and disguises.... [They] march on foot, on horseback, in wagons, carts, coaches, cars, & c. in rich confusion up and down the street wildly shouting, singing, laughing, drumming, fiddling, fifing, and all throwing flour broadcast as they went their reckless way." He noted, "This carnival is permitted by the city authorities, sometimes rather reluctantly, and has been more than once forbidden, as well as the Congo Square dances; but the Creole propensity for these amusements is so strong that their friends are soon placed in power again, and the wild frolics are hailed by acclamation." By the 1850s, crowds of twenty thousand maskers filled the city's streets on Mardi Gras Day. Throughout the antebellum period masking, parading, elaborate floats, and the tossing of sweets to the crowd were important parts of the festivities.[4]

In 1857 a group of prominent Anglo-Americans in the city organized the Mystic Krewe of Comus, and the English-speaking community formally embraced the Carnival tradition. Comus was the first krewe to select a theme for its parade, the first to keep its membership secret, and the first to restrict its parade and masked ball to members only. Before that time, parade organizers had published a time and meeting place in the newspapers and invited all revelers to take to the streets. Comus enabled the city's elites to impose some order on the parades, and drew a distinction between the spectators and the parades that had not existed before. And by parading largely through the American portion of the city, it further announced its Anglo leanings. Comus was in part an Anglo appropriation of the Creole holiday, but it also marked a point of convergence among those two communities.[5]

As the 1781 Spanish restrictions on black revelers suggest, Carnival was never exclusive to the white community, whether Creole or American born. Timothy Flint, a Presbyterian missionary who attempted, with little effect, to plant Protestantism in the southern Babylon in 1823, observed a black festival during Carnival season. He compared "the great Congo dance" to "the 'Saturnalia' of the slaves in ancient Rome." He reported, "Some hundred of Negroes, male and female, follow the king of the wake, who is conspicuous for his youth, size, the whiteness of his eyes, and the blackness of his visage." The king's crown consisted of "a series of oblong, gilt-paper boxes on his head, tapering upward like a pyramid.... By his thousand mountebank tricks, and his contortions of countenance and form, he produces an irresistible effect upon the multitude. All the characters that follow his, of leading estimation, have their own contortions." "Every thing," he marveled, "is license and revelry" as the procession made its way through the streets and blacks and whites watched from the sidelines.[6] He indicated that the African-Creole musicians and dancers usually confined to Congo Square took to the streets during Carnival season. As Flint's description suggests, Carnival as it was practiced by Catholics in the Caribbean and South America, like the Catholic Church and its rituals, provided a cover for Africans in the Americas to retain parts of their African heritage through costuming, dance, and music. The extent to which New Orleans's African-Creole Carnival traditions were influenced by those in other parts of the Americas remains a matter of dispute, but certainly they all share much in common.[7]

Both the Confederate government and Union occupiers prohibited formal parades during the Civil War, and Comus withdrew, but that did nothing to stop New Orleanians from celebrating. The masked balls went on as before, and masked revelers took to the streets as they had traditionally done. In 1863, one surprised Union soldier spotted "a few females of questionable standing" parading in men's attire, and the following year a local newspaper suggested that "Africans" and children would wear costumes, while "crowds of disreputable women in male attire, or in gaudy and tinseled garments of their own wear," would be found "mingling in the throng." When the war ended, Comus once again paraded, and Carnival resumed its former character. The next important innovation came in 1872 with the organization of the Krewe of Rex, intended to enliven the proceedings as a relief for a city worn by war and Reconstruction. A committee was organized by members of the city's elite to organize the new krewe. They proclaimed a king of carnival, Rex, and asked the mayor to turn the city over to him for Mardi Gras Day.

The new king began to issue a series of edicts to close schools, government offices, and businesses for the first daytime parade and, tongue-in-cheek, went on to disperse the state legislature, outlaw taxes, and raise the prices of cotton and sugar to astronomical heights. Rex ordered those living along the parade route to decorate their galleries in his chosen royal colors—green, purple, and gold. The Grand Duke Alexis of Russia announced that he would visit New Orleans during the Carnival season, which influenced the plans. Rex ordered all bands to play his royal anthem, "If Ever I Cease to Love," taken from a play of the time and supposedly made popular by the grand duke's mistress, who was performing it on stage in New Orleans at the time of the duke's arrival. All of this has been interpreted as an homage to the grand duke, though, as historian Reid Mitchell suggests, it was more likely parody. In any event, it was an impressive display with bands, mounted policemen, artillerymen, a fatted ox, hundreds of maskers on horseback and on foot (including whites in blackface and "feminine looking individuals" in carriages who may have been men dressed as women or women dressed as men), and advertising vans that hawked the Gem Saloon, Singer Sewing Machines, Dr. Tichenor's Antiseptic, and other businesses.[8]

One of the most famous Mardi Gras parades on record took place during the 1877 Carnival and celebrated the violent end of Reconstruction in Louisiana and the reestablishment of white rule. The Krewe of Momus took as its theme, "Hades—A Dream of Momus," a vicious satire on the entire Reconstruction era and its local, state, and national leadership, including President U. S. Grant as Beelzebub and General William Tecumseh Sherman as Baal. Other prominent Republicans who came under fire included James G. Blaine and Frederick Douglass. The final float showed the ship of state going up in flames. The satire was so outrageous that some Louisiana Republicans suggested that the organizers should be arrested. If the message were not clear enough, Comus's parade, which closed the season, celebrated the superiority of "The Aryan Race." During the Reconstruction era New Orleans's white supremacists used Mardi Gras parades as a form of political expression and defiance. They did not abandon Carnival and retreat into the private sphere as many white elites did in other parts of the Americas, but rather strengthened their hold on the official celebrations. With white supremacy reestablished in Louisiana and across the South, with northern acquiescence, Carnival became a season of national reconciliation as the city encouraged tourists from across the nation to attend, and parade organizers held out olive branches to their former enemies. During the 1879 celebration, for example, William Tecumseh Sherman was proclaimed "Duke of Louisiana," and U. S. Grant was admitted as a member of Rex. New Orleans successfully used Carnival to promote itself as a tourist destination; the railroads ran special excursions to the city, and a visitor's bureau set up offices across the country from Minnesota to Washington, D.C. to New York City, and tens of thousands of tourists poured into the city.[9]

Between the end of Reconstruction and World War II, Mardi Gras expanded its national appeal as a tourist attraction and at the same time extended its reach into every social sector and racial group. Elite white women organized the first women's non-parading krewe, Les Mysterieuses, in 1900, and others followed. White working-class men organized marching clubs with names like the Jefferson City Buzzards, the Jassy Kids, and the Broadway Swells that essentially went from bar to bar; elite men organized several non-parading krewes, clubs, and debutante societies; and African Americans followed suit with the Illinois Club in 1895. The Krewe of Zulu, the most famous of the black krewes, grew out of a black Carnival marching club called the Tramps, and joined Carnival in 1909. Zulu quickly became notorious for mocking whatever Rex did—if Rex arrived at the riverfront escorted by the U.S. Navy, then the king of Zulu came in a skiff or a tugboat. Instead of bejeweled gowns and regal scepters, Zulu wore blackface, dressed as African natives with grass skirts and spears, and carried a hambone as a scepter. A second black krewe, the Jolly Boys, began parading in 1933. The Mardi Gras Indians made their formal appearance in the late nineteenth century. While their origin is obscure and may date back to the colonial period, the first organized tribe, the Creole Wild West, paraded in 1885, apparently inspired by Buffalo Bill's Wild West Show, which spent the winter of 1884–85 in the city. As more and more tribes sprang up, they developed their own rich traditions—language, music, and elaborate costumes—and the confrontations between them were sometimes violent. Around 1912 a group of black prostitutes organized their own marching club called the Baby Dolls. Dressed in bloomers and bonnets, they paraded through the streets behind the bands. Try as they might, the white establishment could not control Carnival, and everyone claimed the streets— whites of all socioeconomic classes, African Americans, women in masks and costumes, women dressed as men, men dressed as women. As the city worked to attract more tourists to Carnival, its officials also worked to prevent racial incidents that marred the celebrations and sent the wrong message to visitors.[10] In the best Carnival tradition, the lord of misrule continued to hold sway, interrupted only by World Wars I and II.

When Carnival resumed after World War II, it seemed to have lost some of its luster; tourism lagged, hotels were less than half full, and the weekend before Fat Tuesday had no parades at all, creating an unfortunate lull in the festivities. In 1949 Zulu brought national attention to its parade when they invited jazz great Louis Armstrong to serve as King Zulu, a role the New Orleans native relished. Darwin Fenner, captain of Rex, decided that the floats needed to be updated, and he sent Blaine Kern, a young float painter, to Italy and Spain to find inspiration. Kern, who later called himself "Mr. Mardi Gras," introduced large, lavish floats that could be easily adapted to any theme, and launched a hugely successful business that helped define Mardi Gras parades. In 1967 a young advertising executive, Ed Muniz, launched a new krewe called Endymion to parade on the weekend before Mardi Gras. The following year another group with similar ideas met in the home of Adelaide Brennan, one of the city's best-known restaurateurs, to organize a krewe to take the Sunday night slot before Mardi Gras. Calling itself Bacchus, the krewe set out to out-match all others with more elaborate floats, more riders, more throws, and to invite a national celebrity to preside over its festivities. The birth of Endymion and Bacchus, the first super-krewes, gave Carnival a major boost, and their celebrity riders helped bring additional national coverage to the event, which quickly grew into one of the nation's foremost tourist attractions.[11]

By the 1990s Mardi Gras earned its name as the "Greatest Free Show on Earth"; the number of parades increased from twenty-three in 1970 to forty-seven in 2000; it attracted millions of tourists, widespread international media coverage, and brought millions of dollars into the local economy. One of the most explosive challenges to Carnival traditions came in 1991 when the City Council passed an ordinance intended to desegregate parading Carnival krewes. Bitter arguments erupted in a series of hearings, and major krewes like Comus and Momus withdrew from parading rather than comply, even after the ordinance was struck down by federal courts.[12] While the Carnival season always had a reputation for excess and sexual abandonment, and New Orleans has marketed that aspect of Carnival since the late nineteenth century, national media coverage focused increasingly on the rowdier aspects of the festival and helped fuel that behavior. Such coverage entered a new phase in March 1999 when "an army of Playboy.com photographers and producers" descended on Bourbon Street for live coverage of the scene in the French Quarter. Playboy billed itself as "the premier online destination to experience all of Bourbon Street's debauchery, with everything from survival guides to flasher galleries and sexy balcony shows featuring Playmates and models playing up to the crowds." That view of Carnival quickly took hold on the national imagination; in 1998 a *New York Times* article declared, "As cameras for MTV, true-life crime shows and tabloid news programs roll in the French Quarter, the drunken partying has grown so extreme—flashes of nudity have given way to the actual performance of oral sex acts on Bourbon Street—that it is the drunkenness and obscenity itself that threatens to become Carnival's theme." The growth of Mardi Gras and its impact on the local economy, increasing from about $240 million in 1986 to over $1 billion in 2000, showed just how well sex sells.[13]

Carnival is symbolized by the masks of Comedy and Tragedy, and never before did that symbolism appear more apt than in 2006. As James Boyden demonstrates in "Wilt Thou Judge the Bloody City?" above, New Orleanians were stunned when critics from across the nation, particularly from the religious right, blamed the destruction wrought by Katrina and the levee failures on the wickedness of the city and portrayed its destruction as the vengeance of an angry God. A *New Orleans Times-Picayune* editorial summed up this view: "Some voices in Washington are arguing against us. We were foolish, they say. We settled in a place that is lower than the sea. We should have expected to drown. . . . They act as if we are a burden. They act as if we wore our skirts too short and invited trouble."[14] As the chorus of criticism mounted, another editorial in that newspaper asked the same question, "I wonder what New Orleans did to the rest of the country that makes them hate us so?"[15] Many of those critics questioned whether the city should be rebuilt, and even supporters of rebuilding tied federal funding to an insistence that the city change direction. A *Los Angeles Times* editorial reminded its readers of all the city's faults—high poverty and crime rates, a weak economic base, widespread political corruption—and concluded that "New Orleans should take its destruction as an opportunity to change course. . . . [H]aving a strong gay community, lively street culture, great food, tremendous music and lively arts have not been enough."[16] The widespread destruction of the city, including some of its most impoverished but culturally rich areas, made residents fearful of that very sort of change, terrified that the city's historic, rich, and diverse culture might be washed away by the floods. Mayor Ray Nagin summed up the dilemma: "It's a two-edged sword," he said. Going ahead with Carnival would "send out the signal that New Orleans is not dead, that we've honored our tradition of 150 years. . . . But it also sends out the signal that we're OK, and 'There they go again, partying when they have serious challenges.'"[17] As one local news outlet reported, "the question sprang almost instantly from the lips of reporters and pundits nationwide: Will Mardi Gras roll? Will New Orleanians dare party in the ruins?"[18]

A part of the problem reflected in the criticism lies in how we think of natural disasters and their impact. Disaster response has largely been grounded in a view of society as a collection of individuals, and individuals and society are often seen as two separate systems. Social scientists have studied the effects of natural and man-made disasters on individuals (posttraumatic stress and grief, for example), and on societies (the breakdown of law and order, economic devastation, for instance), but between these two levels lie a range of "mid-level" relationships and structures that are also disrupted by disasters, but which have not been studied in depth. These "small groups," whether they are churches, social clubs, sports clubs, or other collections of like-minded people, have an important role to play in disaster recovery. The loss of these groups can result in a lack of intimacy and social support, a reduction in caring behavior, a breakdown in community communication structures that can contribute to individual isolation. The destruction of these small groups can contribute to "culture loss," and scholars have found that the sense of loss is especially pronounced in the wake of natural and environmental disasters that severely damage or destroy the lived environment.[19] New Orleans is famous for its distinctive culture, and the city's residents treasure its peculiarities and embrace its rich cultural diversity that has so far withstood the homogenization that has erased so much regional distinctiveness in America. One of the factors contributing to that cultural resilience is the stability of the city's neighborhoods. While the population of the United States is a mobile one, New Orleanians are deeply rooted. A 2006 study found that, while 79 percent of Americans lived in the same county or parish they were born in, that figure rose to 85 percent in New Orleans, and in some neighborhoods like the Lower Ninth Ward, that figure rose to 97 percent.[20] That tenacious dedication to the city, its neighborhoods, and its traditional way of life was clearly threatened by the devastation and depopulation of huge swaths of the city.

The flooding and destruction of large parts of New Orleans can be compared to other natural and man-made disasters with a similar impact on the lived environment. The destruction caused by the 1989 *Exxon Valdez* oil spill in Alaska had a devastating impact on the indigenous population and their culture, but the courts refused to accept that the disaster deprived them of that resource since culture, the court ruled, was embedded in hearts and minds and not in the physical landscape. The courts failed to see the importance of landscapes and the lived environment, of sacred spaces, of the deeply rooted stories that tie individuals to their ancestors, and the dangers posed by the loss of place and the "dissonance between what once existed and what [now] exists." In this context, disaster researchers are growing more sensitive to the importance of neighborhoods as sites where "meaningful social action can be both generated and interpreted."

It is vital to understand that neighborhoods can be destroyed, and with them goes the deep accumulation of experience and identity that are so crucial to a larger sense of belonging and community. Scholars of disaster response and recovery are also recognizing the "multifunctional nature of large cities." It is essential to recognize, as some scholars have, that New Orleans has nurtured a unique and deeply rooted culture. James K. Mitchell, professor of geography at Rutgers University and chair of the International Geographical Union's Study Group of the Disaster Vulnerability of Megacities, noted that New Orleans has served "as both a shelter and a muse. . . . Vernacular lifestyles," he observed, "are confirmed, expressed, and renegotiated through performances, sometimes ritualized (e.g., Mardi Gras, second lining), but often more casual. . . . More than most others, this is a city that has tested the limits of social control and cherishes an ebullient, wayward self-image."[21] Students of disaster have asked why people would consider returning to a place that remained vulnerable and where they had suffered great loss. A part of the suffering caused by a megadisaster like Katrina is the severe disruption of people's lives, the loss of the cohesiveness of life as they knew it, and the erosion of the meanings that were the foundation of people's sense of place, community, and identity. Scholars have found that the relationship people have with their chosen environment is a source of healing, and redeveloping and regaining a sense of place plays an instrumental role in that process.[22]

For New Orleanians, the question of whether to hold Carnival was complex. Some natives, particularly those displaced, condemned the decision to go ahead with Mardi Gras in the face of the death and destruction that the hurricane inflicted on the city. Samuel Spears, displaced to Houston, agreed with the critics. "With them putting on Mardi Gras, without still having not addressed the basic human needs in this city, why that's just a slap in the face. I can't go home, but they can have a parade? That's ridiculous," Spears said.[23] ChiQuita Simms, a New Orleans native displaced to Atlanta, spoke for many refugees when she criticized the decision of city leaders to go ahead with the festivities: "We're not against Mardi Gras, we're against their priorities." Simms organized a protest by displaced New Orleanians in Atlanta to coincide with a Saints football game there. Another refugee, Jerome Casey, agreed, "They shouldn't be preparing for Mardi Gras. . . . They should be trying to get New Orleans back on its feet." Simms and Casey, along with many other critics, saw a racial element in the decision. As Casey put it, "In New Orleans, everything is about race at the end of the day."[24] Ernest Johnson, president of the Louisiana branch of the NAACP, explained, "Eighty percent of those whose homes were destroyed were African-American, while 80 percent of the people who are going to do Mardi Gras are white. You have black folks who are still out of the city and can't come back to their homes, and you have white people who want to have a party. You have to draw the conclusion that this is a racial divide."[25]

But most locals took a different view of the celebration that so defines the city's cultural life. Many people shared the defiant attitude of Louise Maloney, who vowed, "I'll get a red wagon, fill it with beads and walk down the street. I don't give a s[hit] what anybody else does. I'm having . . . Mardi Gras."[26] Former mayor Marc Morial, president of the National Urban League, agreed with Maloney. In an interview before the National Press Club, Morial said, "I have been torn about Mardi Gras—very, very torn. But . . . you know, you really can't cancel Mardi Gras. Because it's in the DNA. When I say you can't cancel—you can cancel it, but it wouldn't be canceled. . . . [P]eople would go out, and they would find a way to party on Mardi Gras day even if there wasn't a single parade on the street. That's the truth."[27]

As the debate continued, it became increasingly clear that Morial was right—Mardi Gras is in the DNA of the city. What is perhaps not clear to outsiders is the degree to which Mardi Gras mobilizes such a large part of the city's residents, often through the mid-level organizations like the krewes, marching bands, social aid and pleasure clubs, and other organizations that together make the event happen. Mardi Gras is not organized by the city government or funded by it, nor does it have corporate sponsors or advertisers. It is a massive celebration the city throws for itself. Even when the city decided to go ahead with Mardi Gras celebrations in 2006, it was unclear how many of the city's krewes would be able to participate. Lloyd Frischhertz, one of the fraternity brothers who founded the Krewe of Tucks in 1969, said, "I did not want to parade. I thought it would be an insult to everybody who lost their home, had to move, lost somebody." But as Frischhertz worked to clean his own flooded home, his phone began to ring. Captains from other krewes called to say they wanted to roll. Frischhertz and the others were swayed by arguments that pulling the plug on the city's most symbolic and cherished event would send a message that the city had given up. As former Tucks King Rudy Ormond put it, "If you're not gonna have Mardi Gras, why have Christmas? Why have New Year's?" City officials and krewe captains met and agreed on a shorter season and a more limited parade route. In the end, twenty-eight krewes prepared to roll, compared to thirty-four the year before.[28]

The 2006 parade marked the 150th anniversary of Mardi Gras, and after all the tragedy and doubts, New Orleanians embraced it. Satire and challenges to authority have always been a feature of carnivals, and so it was in 2006. The Krewe de Vieux, which opens the season and is the only parade to wind its way through the narrow streets of the French Quarter, set the tone. Always noted for its satire, the krewe chose as its theme "C'est Levee," a play on the

French phrase "such is life." One float pleaded "Buy Us Back Chirac," a nod to the federal government's slow response to rebuilding the city. Members of the Knights of MONDU wore individually decorated refrigerator headpieces, a reference that only those of us who cleaned rotting food from our refrigerators could fully appreciate. The Krewe of Mama Roux presented "Home is Where the Tarp Is," its members costumed in the blue material that protected damaged roofs throughout the city. According to the krewe's newsletter, they got a special discount on the blue material from "a FEMA sub-sub-sub-sub-sub contractor for a mere $985 per yard." The Krewe of K.A.O.S. (Kommittee for the Aggravation of Organized Society) took aim at FEMA director Michael Brown with their theme "K.A.O.S. Rules FEMA." Its float carried an empty throne with a sign announcing that Brown, anointed as grand marshal by the krewe, was "out to dinner," while other signs on the float promised that decorations and beads were "on the way."[29] It was clear that Mardi Gras 2006 would be unlike any other. Orleans Parish coroner Dr. Frank Minyard, who helped identify many of the 1,103 Louisiana victims who perished in the hurricane, was a grand marshal of Saturday's parade, happily tossing beads from one of the floats.[30]

Perhaps the most vicious satire of the entire 2006 season came from the Knights of Chaos, who choose the theme "Hades—A Dream of Chaos." They threw cards that proclaimed "This tableau could be called 'Reconstruction II,'" a reference to Mardi Gras 1877 when the krewe of Momus mocked the national and state Republican leadership with its theme, "Hades—A Dream of Momus." The Knights of Chaos took inspiration from the 1877 parody. Their floats, many patterned directly after those of 1877, portrayed terrible scenes from hell with themes like "The Headless State," "Carpetbaggers," "Homeland Insecurity," and "The Corpse of Engineers." On a float called "The Inferno," Governor Kathleen Blanco, Mayor Ray Nagin, and Michael Brown were portrayed as "infernal cooks stirring a huge cauldron of human gumbo in the Superdome as members of Congress forced people into the boiling pot with pitchforks, and a leering George W. Bush presided over the whole scene as the horned Satan incarnate."[31]

Especially popular are the local high-school marching bands that march behind every float. With many schools closed and students dispersed, many school bands were unavailable. Three historically African American Catholic schools—St. Augustine, St. Mary's, and Xavier Prep—responded by merging their depleted bands into the Max Band, made up of about a hundred students. Many of the students stayed with friends or relatives in order to participate, and their parents drove in to see them march during Carnival. Band director Lester Wilson noted, "For the kids, it's an opportunity to prove they can overcome adversity. You've got kids coming from Baton Rouge just to go to school, and staying late for band practice—that's commitment." The band members wore special gold uniforms, the only color shared by the three schools, with the collars and cuffs carrying the unique colors of each school. Gregory Malone, a seventeen-year-old St. Augs student who traveled back and forth to Houston each weekend to be with his displaced family, said, "I get discouraged sometimes, but my peers cheer me up. . . . I've worked hard for four years, so you have to march. It feels good to put on that uniform, and wear it with pride."[32]

There were serious moments, too. Members of the Zulu Aid and Pleasure Club held a traditional jazz funeral for those who died in the storm, and before their parade rolled, members gathered at their headquarters and lit ten candles for members who had died in the storm, and an eleventh candle in memory of all other victims.[33] The last float in the Krewe of Muses parade was dedicated to Mnemosyne, the goddess of memory and mother of the Muses. The float was empty; its banner read, "We Celebrate Life, We Mourn the Past, We Shall Never Forget."[34] Many residents stood along the parade route holding hand-printed signs with street names—Memphis, Louis XIV, Fleur de Lis. "These people were holding cardboard signs with the names of their flooded streets. . . . It wasn't like they were asking for beads. It was a reminder," one reporter noted.[35]

Of all the many distinctive groups that take to the streets on Mardi Gras Day, none are more distinctive or more central to the city's multicultural heritage than the Mardi Gras Indians. Like many Mardi Gras institutions, their origins are shrouded in mystery. Legend has it that their origins date back to the early eighteenth century when African slaves joined Native Americans in a bloody rebellion against French rule, and some Native Americans assisted runaway slaves in the early history of Louisiana. The 1781 Spanish regulations against slaves wearing feathers during Mardi Gras may well offer the earliest evidence of the Indian tradition. The Indians made their first recorded appearance in 1885, when the Creole Wild West marched in a parade, and before Katrina struck, about twenty-five tribes existed. The Indians are most famous for their elaborate costumes, each carefully sown by hand by the Indian who wears the suit. Richly decorated with colorful feathers and beadwork, the costumes may take a full year to sew, cost thousands of dollars, and weigh over a hundred pounds. The Indians' chants predate jazz and have influenced musicians from Jelly Roll Morton to the Neville Brothers. Donald Harrison, a noted jazz musician and Big Chief of the Guardians of the Flame, observed, "A large percentage of American popular music, and jazz music, has been influenced by this root culture. It's the link for all the music that we hear, that takes us right back to . . . Africa."[36]

Many of the Indians lived in working-class neighborhoods devastated by Katrina, and they were scattered by the storm. Indians in the Lower Ninth Ward saw a lifetime of elaborate and valuable suits ruined by floodwaters. Even when they managed to return home, they had to choose between spending their meager funds on their suits or on food and home repairs. Norman Cook, a member of the Creole Wild West, said, "You have to ask yourself: Should I spend $2 on beads or try to buy something to eat?" Joyce Marie Jackson, an ethnomusicologist who has studied the Indian tribes, noted that displacement took a heavy toll on the members: "The ones who are away, outside of New Orleans, are suffering an identity crisis. . . . In New Orleans, they were a Big Chief or a Gang Flag. In Atlanta or Houston, they're nobody."[37]

Determined to preserve their rich cultural heritage, many of the Indians took to the streets on Mardi Gras Day as their members returned from around the country to join in the annual tradition. Big Chief Larry Bannick of the Golden Star Hunters explained why he chose to mask: "The city is tore up. The city is tore up like a bomb hit it. A lot of people don't want to mask. I do it because I love it. I do it because it's my joy and my pain. When you hit that tambourine that morning and this . . . old lady come down the street and say, baby, you're pretty, that's your glory. All them little children holler, look at my big chief. My glory is my neighborhood. . . . I haven't cried since the storm. I've tried, but I can't cry. Maybe when I walk through this city Mardi Gras day I might cry, because my city is tore up. But we still coming. People say y'all shouldn't, but we coming. We may not make a big show, but when the history books write the story of 2006, they going to say the Mardi Gras Indians played their part."[38] Alphonse "Dowee" Robair, a member of the Red Hawk Hunters, led his tribe on an emotional march through their ruined neighborhood. Symbolically, they began their march in prayer at the foot of the Claiborne Bridge over the Industrial Canal near the breach that wiped out the Lower Ninth Ward. Robair said, "It's about keeping up tradition. I'm going to do what was taught to me as a child."[39]

David Simon, producer of the HBO drama *Treme* set in post-Katrina New Orleans, tried to capture the significance of the Mardi Gras Indians and the culture that produced them. He observed, "There's a thing about being capable of a great moment. . . . This city is capable of moments unlike any moment you'll ever experience in life. To see an Indian come down the street in full regalia on St. Joseph's Night on an unlit street of messed-up shotgun houses and one burned-out car, and he's the most beautiful thing on the planet, and everything else around him is falling down. It's a glorious instant of human endeavor." And according to Simon, it's a moment that can only be experienced in New Orleans. "Lots of American places used to make things," he said, "Detroit used to make cars. Baltimore used to make steel and ships. New Orleans still makes something. It makes moments. . . . [T]hey can take a moment and make it into something so transcendent that you're not quite sure that it happened or that you were a part of it."[40]

Or begin the day outside of the Quarter in the Marigny neighborhood on Frenchmen Street, where the Society of St. Anne assembles, regarded as "arguably Carnival's most colorful semi-organized display of New Orleans' flamboyant eccentricity." Predominantly gay, the outlandishly costumed revelers, led by the Storyville Stompers brass band, march through the Quarter to Canal Street and take their place behind the Rex parade. But in keeping with Carnival's double face, their bawdy parade ends on a solemn note. Their walk ends at the Mississippi River for a "baptismal ceremony." The leaders dip their ribbon-covered hula hoops—which they call crab nets—into the muddy water and sprinkle the drops over the crowd. The Storyville Stompers play hymns like "Down by the Riverside" and "I'll Fly Away" while the leaders scatter the ashes of members and friends who have died during the past year into the swirling muddy waters, a tradition they began in the 1980s as the AIDS epidemic decimated their numbers.[41]

In keeping with 150 years of tradition, New Orleanians donned costumes and took to the streets on Mardi Gras Day, almost all of them taking aim at Katrina. Palmer Stubbs proudly stood beside his "Katrina Deli cart" with selections such as "Sheet-Rock Candy" and "Oysters Hepatitis B-ienville." As Mitchell Gaudent pointed out from inside a giant fleur-de-lis with a screw through it, finding satire in 2006 was "like shooting fish in a barrel . . . and people are embracing me."[42] Over and over again, parents said that they wanted to see Carnival go forward for the sake of the children. It may seem incongruous to outsiders that a festival perceived to be one characterized by debauched excess is considered by locals to be a family affair. Antoinette Butler, who lost her home in New Orleans East, gestured toward her three children, ages one, three, and ten. "We're here for them," she observed. Cindy Pierce, a lifelong resident, pointed to her two daughters standing on the parade route wrapped in red tape with slogans like "Where's my FEMA trailer." "I'm proud of them," she said, "They don't dress like this every day, but this is Mardi Gras. It's ours. It's special. We own it." The story of the Shultz family is representative of how many extended families spend Mardi Gras Day. Every year over a dozen family members gather at the corner of St. Charles Avenue and Marengo Street, and family members who were displaced to Texas returned to the spot in 2006. As Robert Shultz Jr. remarked, "That's the key to it, the annual celebration with family. I haven't seen a lot of friends and family since the hurricane hit, haven't had a chance to give them a hug." Some displaced residents drove in from as far away as Atlanta or Dallas just for the day.[43]

And these are only a few examples of the many-layered meaning of Carnival for the people of New Orleans. Anthropologists understand group events like Mardi Gras as *functional.* We are social animals, and such rituals renew the bonds that hold a community together. In that sense, every Mardi Gras float that rolls, every Mardi Gras Indian tribe that parades, every jazz band that takes the stage, every high-school band that marches represents a reknitting of the city's frayed social fabric. Noted anthropologist Victor Turner saw the collective ecstasy displayed in public rituals as an expression of *communitas,* evidence of the love and solidarity that can arise within a community. Such occasional sources of relief as ecstatic or unruly group behaviors serve a legitimate purpose in society by preventing social structures from becoming rigid and unstable. Anthropologist Roger Abraham explained, "The vocabulary of festival is the language of extreme experiences through contrasts.... The body is made into an object of dressing up, costuming, and masking.... And, of course, singing and dancing and other kinds of play are part and parcel of festive celebrations, again with the idea of overextending the self. All of these motives underscore the spirit of increase, of stretching life to the fullest that lies at the heart of festive celebrations."[44] Herein lies the deeper significance of Mardi Gras for a devastated city, and why it was so essential to revive it in the wake of the death and destruction and loss of community that accompanied Katrina.

Louise Maloney, who vowed to have Mardi Gras if it meant filling her little red wagon with beads and pulling it down St. Charles Avenue, wept as she watched the first parade of the season; "this is what we do," she said. "We take tragedy and make it into beauty and hilarity. And we're also showing pride in being New Orleanians and expressing ourselves like never before."[45] When locals worked to revive Mardi Gras after the disruptions of World War II, Robert Tallant, author of a 1947 history of Carnival, wrote, "If there is any world left in which human beings still laugh and still, even on rare occasions, have fun, there will be a Mardi Gras. . . . It will live through whatever catastrophes occur. . . . Men cease to laugh only when they are very ill or when they have become beasts. . . . That is why Mardi Gras is not a trivial matter but a very important one."[46] Mardi Gras 2006 brought smiles and laughter and tears. It restored the soul of a devastated city. A commentary in a local magazine summed up the feelings of the people of New Orleans: "So, if anyone asks, 'Why have Mardi Gras' tell them the answer is simple: Because it matters. It is how we heal, how we deal with whatever life throws at us. Now more than ever, we need to show the world that we are healing, and that we will not let tragedy take our soul, destroy our culture, or break our spirit."[47]

Notes

1. "Why Mardi Gras Matters," *Gambit Weekly,* January 17, 2006.

2. James K. Mitchell, "The Primacy of Partnership: Scoping a New National Disaster Recovery Policy," *Annals of the American Academy of Political and Social Science* 604 (March 2006): 244–47; Anthony Oliver-Smith, "Anthropological Research on Hazards and Disasters," *Annual Review of Anthropology* 25 (1996): 304, 309, 311, 318, 321–22.

3. Barbara Ehrenreich, *Dancing in the Streets: A History of Collective Joy* (New York: Metropolitan Books, 2006), 87–95, 147–52, first quote on 88; Mikhail Bakhtin, *Rabelais and His World* (Bloomington: Indiana University Press, 1941); Natalie Zemon Davis, "Women on Top," in *Society and Culture in Early Modern France: Eight Essays by Natalie Zemon Davis* (Stanford, Calif., 1975), 124–51; Edward Muir, *Ritual in Early Modern Europe* (Cambridge, U.K.: Cambridge University Press, 2005); Sophie White, "Massacre, Mardi Gras, and Torture in Early New Orleans," *William and Mary Quarterly,* 3rd ser., vol. 70 (July 2013): 497–538; Reid Mitchell, *All on a Mardi Gras Day: Episodes in the History of New Orleans Carnival* (Cambridge, Mass.: Harvard University Press, 1999), 17–20.

4. James Gill, *Lords of Misrule: Mardi Gras and the Politics of Race in New Orleans* (Jackson: University Press of Mississippi, 1997), 27–35; James R. Creecy, *Scenes in the South, and Other Miscellaneous Pieces* (Washington, D.C.: Thomas McGill, 1860), 43–45; Mitchell, *All on a Mardi Gras Day,* 10–28; Perry Young, *The Mystic Krewe: Chronicles of Comus and his Kin* (New Orleans: Louisiana Heritage Press, 1969), 35–76; Samuel Kinser, *Carnival, American Style: Mardi Gras at New Orleans and Mobile* (Chicago: University of Chicago Press, 1990).

5. Mitchell, *All on a Mardi Gras Day,* 23–28; Young, *The Mystic Krewe,* 65–76.

6. Timothy Flint, *Recollections of the Last Ten Years, Passed in Occasional Residences and Journeyings in the Valley of the Mississippi, from Pittsburg and the Missouri to the Gulf of Mexico, and from Florida to the Spanish Frontier . . .* (Boston: Cummings, Hilliard, and Co., 1826), 140.

7. Ehrenreich, *Dancing in the Streets,* 164–69; Mitchell, *All on a Mardi Gras Day,* 29–37.

8. Mitchell, *All on a Mardi Gras Day,* 51–64, quote on 58; Gill, *Lords of Misrule,* 59–60, 93–100.

9. Mitchell, *All on a Mardi Gras Day,* 82–95; Young, *The Mystic Krewe,* 92, 159; J. Mark Souther, *New Orleans on Parade: Tourism and the Transformation of the Crescent City* (Baton Rouge: Loui-

siana State University Press, 2006), 132–58. On publicity for railroads, see Illinois Central Railroad Company, *New Orleans for the Tourist: The Southern Metropolis Replete with Interesting Evidences of the Old French and Spanish Civilization, . . . the City of Social Brilliancy and the Home of the Mardi Gras* (Chicago: Illinois Central R.R., c. 1896), and for steamboats, Pittsburgh and Cincinnati Packet Line, *Mardi Gras Excursions* (Pittsburgh: A. J. Henderson, 1913).

    10. Mitchell, *All on a Mardi Gras Day,* 113–28, 131–41; Gill, *Lords of Misrule,* 140–42, 157–59, 165, 170–72; Anthony J. Stanonis, *Creating the Big Easy: New Orleans and the Emergence of Modern Tourism, 1918–1945* (Athens: University of Georgia Press, 2006), 15–21, 170–94; "Rally of the Dolls," *Gambit,* February 16, 2009; "He's the Prettiest: A Tribute to Big Chief Allison 'Tootie' Montana's 50 Years of Mardi Gras Indian Suiting," *Louisiana's Living Traditions,* www.louisianafolklife.org/LT /Virtual_Books/Hes_Prettiest/hes_the_prettiest_tootie_mont ana.html.

    11. Gill, *Lords of Misrule,* 200.

    12. Mitchell, *All on a Mardi Gras Day,* 192–201, Gill, *Lords of Misrule,* 221–62.

    13. See www.playboy.com/worldofplayboy/features/10th_anniversary/ (first quote); New York *Times,* February 23, 1998 (second quote); Kevin Fox Gotham, "Marketing Mardi Gras: Commodification, Spectacle, and the Political Economy of Tourism in New Orleans," *Urban Studies* 39 (2002), 1745–46.

    14. *New Orleans Times-Picayune,* November 20, 2005.

    15. Ibid., November 22, 2005.

    16. *Los Angeles Times,* September 4, 2005.

    17. Brian Thevenot, "Their Mardi Gras . . . Our Mardi Gras," NOLA.com, February 19, 2006.

    18. Ibid.

    19. Craig Higson-Smith, "'Linking' and 'Empowering': Key Concepts for Intervention Following War and Disaster," *Development in Practice* 9 (May 1999): 333–37; Stuart Kirsch, "Environmental Disaster, 'Culture Loss,' and the Law," *Current Anthropology* 42 (April 2001): 167–98.

    20. Peter Wagner and Susan Edwards, "New Orleans by the Numbers," *Dollars & Sense,* www .dollarsandsense.org/archives/2006/0306wagneredwards.html (accessed May 9, 2011).

    21. Kirsch, "Environmental Disaster," 175–76, first and second quotes on 175; Mitchell, "Primacy of Partnership," 228–55, quotes on 244.

    22. Helen M. Cox and Colin A. Holmes, "Loss, Healing, and the Power of Place," *Human Studies* 23 (January 2000): 63, 67–69, 74.

    23. *Chicago Tribune,* February 14, 2006.

    24. Ibid., December 7, 2005; Political Transcript Wire, February 14, 2006.

    25. *Dallas Morning News,* February 8, 2006.

    26. *Times-Picayune,* 19 February, 2006.

    27. National Press Club Luncheon with Marc Morial, president, National Urban League, February 14, 2006, press.org/sites/default/files/021406_morial.pdf.

    28. *Dallas Morning News,* February 8, 2006.

    29. "A Ghostly Mardi Gras," Salon.com, February 25, 2006.

    30. *Chicago Tribune,* February 19, 2006.

    31. Thevenot, "Their Mardi Gras . . . Our Mardi Gras."

    32. Ibid.

    33. "Mardi Gras Group Holds Somber Katrina Remembrance," blog.nola.com/mardigras/2006 /02/mardi_gras_group_holds_somber.html.

    34. Krewe of Muses, www.kreweofmuses.org/archives/2006_narrative.html.

35. "Mardi Gras Is Long Gone . . . ," NOLA.com, March 8, 2006.

36. *Chicago Tribune,* December 26, 2006. On the Mardi Gras Indians, see Mitchell, *All On a Mardi Gras Day,* 113–25; "Behind the Lines: The Black Mardi Gras Indians and the New Orleans Second Line," *Black Music Research Journal* 14 (Spring 1994): 43–73.

37. *Chicago Tribune,* December 26, 2006.

38. "Big Chiefs Continue Mardi Gras Indian Tradition," *All Things Considered,* National Public Radio, February 28, 2006.

39. Thevenot, "Their Mardi Gras . . . Our Mardi Gras."

40. "The HBO Auteur," *New York Times Magazine,* March 15, 2010.

41. "Mardi Gras in New Orleans: Before the St. Anne Procession," gayhighwaymen.wordpress. com/2011/03/12/mardi-gras-in-new-orleans-before-the-krewe-of-st-anne-procession/. For more information on the St. Anne Society, see wn.com/Mardi_Gras_in_New_Orleans_Society_of_Saint _Anne's_Parade.

42. RayK, Rolling with the Punches," March 1, 2006, www.nola.com/festivals/index.ssf/2006 /03/index_2.html.

43. Thevenot, "Their Mardi Gras . . . Our Mardi Gras."

44. Roger D. Abraham, "The Language of Festivals: Celebrating the Economy," in *Celebration: Studies in Festivity and Ritual,* ed. Victor Turner (Washington, D.C.: Smithsonian Institution Press, 1982), 167–68.

45. Thevenot, "Their Mardi Gras . . . Our Mardi Gras."

46. Robert Tallant, *Mardi Gras . . . as It Was* (Gretna, La.: Pelican Publishing Co., 1989), x.

47. "Why Mardi Gras Matters."

CONTRIBUTORS

Thomas Jessen Adams is lecturer in history and American studies at the University of Sydney. He was American Council of Learned Societies New Faculty Fellow and Andrew Mellon Postdoctoral Fellow at Tulane University. His research focuses on a variety of topics in U.S. history, including political economy and labor, urban history, social movements, legal history, and race and gender. Recently he has begun to examine the contemporary history, politics, and culture of New Orleans and the Gulf South.

James Boyden is associate professor of history at Tulane University. He specializes in early modern Spain, the Renaissance, and court culture. His publications include *The Courtier and the King: Ruy Gómez de Silva, Philip II, and the Court of Spain* (1995).

Richard Campanella, a geographer with the Tulane School of Architecture, is the author of seven critically acclaimed books, including *Geographies of New Orleans: Urban Fabrics before the Storm* (2006) and *Bienville's Dilemma: A Historical Geography of New Orleans* (2008), both winners of the Louisiana Endowment for the Humanities Book of the Year Award, and *Lincoln in New Orleans* (2010), winner of the Williams Prize for Louisiana History. He has also published numerous articles on historical geography, geographic information systems (GIS), and remote sensing.

Andrew Diamond is professor of American history and civilization at the Université Paris Sorbonne, where he co-directs the research center Monde Anglophone: Politiques et sociétés (MAPS). His research has focused on race, political culture, and inequality in the metropolitan United States. His publications include *Mean Streets: Chicago Youths and the Everyday Struggle for Empowerment in the Multiracial City, 1908–1969* (2009) and *Histoire de Chicago* (2013).

Romain Huret, professor at the Ecole des haute études en sciences sociales in Paris, is the author of several books, most recently *American Tax Resisters* (2014). His current book project, focusing on the trial of Andrew W. Mellon, is to be published in 2015.

Jean Kempf is professor of American studies at the University of Lyon 2. He specializes in the history of photography—especially documentary photography—as well as memory studies.

Sara Le Menestrel is an anthropologist and a research fellow at the National Center for Scientific Research (CNRS), affiliated with the Center for North American Studies at the École des Hautes Études en Sciences Sociales (EHESS) in Paris. Her research interests include the role of music in managing differences. Her book *Negotiating Difference in French Louisiana Music: Categories, Stereotypes, and Identifications* is forthcoming from the University Press of Mississippi. Since 2005, she has extended her research interests to the anthropology of disaster through post-Katrina and post-Rita Louisiana.

Anne M. Lovell is senior research fellow emeritus at the French national health research institute (INSERM) and Cermes3 (University of Paris Descartes). Her current work focuses on madness, disaster, and the politics of time. Her research interests include addiction, pharmaceuticals and globalization, comparative psychiatries, and the social and epistemological history of psychiatric epidemiology. In 2013, she co-wrote *Face aux désastres: Une conversation à quatre voix sur la folie, le care, et les grandes détresses collectives*.

Bruce Boyd Raeburn researches and teaches on jazz, particularly in New Orleans. He is curator of the Hogan Jazz Archives and director of Special Collections at Tulane University's Howard-Tilton Memorial Library. His publications include *New Orleans Style and the Writing of American Jazz History* (2009).

Randy J. Sparks is professor of history at Tulane University, where he specializes in southern history, the early modern Atlantic world, and American religious history. His most recent book is *Where the Negroes Are Masters: An African Port in the Era of the Slave Trade* (2014).